The Hare and the Tortoise:
Clean Air Policies in the United States and Sweden

The Hare and the Tortoise: Clean Air Policies in the United States and Sweden

Lennart J. Lundqvist

Ann Arbor
The University of Michigan Press

Copyright © by The University of Michigan 1980
All rights reserved
Published in the United States of America by
The University of Michigan Press and simultaneously
in Rexdale, Canada, by John Wiley & Sons Canada, Limited
Manufactured in the United States of America

Library of Congress Cataloging in Publication Data
Lundqvist, Lennart, 1939–
    The hare and the tortoise.

    Bibliography: p.
    Includes index.
    1. Air—Pollution—United States. 2. Air—Pollution—Sweden. 3. Air—Pollution—Law and legislation—United States. 4. Air—Pollution—Law and legislation—Sweden. I. Title.
  TD883.2.L86   353.008′232   79-19592
  ISBN 0-472-09310-X

*To Maude, Åsa, and Anja
who know what it means
and what it has meant*

Preface

Planning for this comparative study of environmental policy began many years ago. But planning such a study is one thing; finishing it turned out to be a totally different enterprise.

In fact, the story of this book resembles Aesop's fable. Like the hare, I started out with high hopes and great speed. Predictably, I soon ran into trouble. I found myself overlooking too many roadblocks, warning signs, and dangerous spots along the road to meaningful comparison. It certainly seemed that this study was taking me wherever carrots were to be found. When goals and deadlines began haunting me, I had to change style completely. I confined myself to the speed of the tortoise and also tried to adopt her more methodical, step-by-step way of approaching goals. It is, of course, the reader's privilege to judge how well I have succeeded in these efforts.

The result is this comparative study of clean air policy development and change in the United States and Sweden. Like all studies carrying the *policy* brand, this one has to be placed in a proper context. The rapid increase in public policy studies has caused much concern and confusion with regard to the theoretical, methodological, and conceptual framework of such studies. People talk about policy studies, policy analysis, and policy evaluation. Whether or not they mean the same thing is not always clear to the audience. Some think the proper theoretical focus should be on systems characteristics, systems performance, and policy. Others emphasize the necessity to look at public policy from the perspective of rational choice. Some put their emphasis on quantitative and correlational analysis, while others opt for in-depth analyses of the motivations, intentions, and cognitions of political actors. What causality is to the former, contrivance and rationality is to the latter.

As a *policy study*, this is an ex post facto effort to describe and explain how and why clean air policies have developed and changed in the United States and Sweden. Unlike policy analysis proper, it is not concerned with the merits and drawbacks of the policy alternatives per se in terms of goal achievement and effectiveness. It is concerned with the relationship between policy content and the cultural, economic, social, and political factors usually believed to influence policy content. The key question for most comparative policy studies so far has been whether differences or similarities in these background factors result in differences or similarities in policy content.

Most comparative policy studies have tried to assess this relationship without looking at policymakers' behavior and actions. However, background factors do not make policy. Policymakers do. The interesting thing in policy explanation is what policymakers actually make of these background factors. Thus I try to explain policy content in terms of rational choice. I try to show how those in positions of authority, as well as those advocating alternatives, go about defining what they think is an appropriate *and* feasible response to a given set of circumstances. The underlying assumption is that policymakers, whatever system they are working in, generally avoid alternatives perceived to have negative substantial and strategic consequences. Policymakers' perceptions and interpretations of a given set of circumstances in each country and their intentions and strategies are used to explain the content of clean air policy.

The book consists of three parts. Part one, The Setting, contains a description and a comparison of the major clean air policy choices made in the two countries in 1969–70. It also contains a discussion of the theoretical and conceptual framework of the study, and poses questions to be answered in the rest of the book.

Part two, The Choice, analyzes the intentions, motivations, and calculations of the policymakers and other relevant actors at the time of policy choice. It looks at the arguments for or against the chosen major control approaches (chapter 3), the allocation of policy authority and responsibility (chapter 4), and the mechanisms for public participation in policy implementation (chapter 5). Chapter 6 pulls it all together in an effort to explain the policy differences. The major conclusion is that politics means more than environment and technology.

Part three, The Change, traces the changes in clean air policies during the 1970s, with special emphasis on what happened when policymakers became preoccupied with energy policy. The final chapter examines the theme emerging from the study, i.e., that the two distinct patterns of policy choice and change are best explained in terms of policymakers' rational assessment of the *political* opportunities and obstacles they perceived as relevant to clean air policy choice in their respective systems. A limited effort is also made to find out which of the patterns has been most successful so far, the United States "hare" or the Swedish "tortoise."

Help and assistance of many kinds from many sources were invaluable to me in making this study. Among the many colleagues who have been involved, I first acknowledge Keith Caldwell, whose contribution I value above all others. I am also most indebted and grateful to Marv Olsen and Tim Tilton for their advice and friendship. Bert Hanson offered helpful comments during his year in Sweden. He and Steve Kelman commented on early drafts. Tom Anton and Joe Board provided support and encouragement at an opportune moment. Aaron Wildavsky pointed out that contemporary political reality is more than ancient fables.

Most of this book was written at the Department of Government of Uppsala University. I am most indebted to all of my colleagues at *Skytteanum*. Their incisive questioning and eloquent criticism provided a most stimulating research *milieu*. Special thanks are due to Stefan Björklund, who helped and encouraged me at a time when I really needed it. Sverker Gustavsson and Evert Vedung commented upon early drafts, and Olof Petersson provided unfailing support and encouragement.

All of these people contributed something to my thinking about this study. Needless to say, I alone am responsible for whatever use I have made of their comments.

This study would not have been possible without the generous financial assistance of the American-Scandinavian Foundation's Bernadotte Fund, the Swedish Social Science Research Council, and the Helge Ax:son Johnsson Foundation. That assistance made it possible to spend one year (1972–73) at Indiana University. I am most grateful for the help and assistance provided by that university's Department of Political Science and School of Public and Environmental Affairs.

Maude's, Åsa's, and Anja's contributions, of course, go far beyond collegial encouragement and enthusiasm. Their love, affection, faith, and patience made it all possible.

Lennart J. Lundqvist
Uppsala, Sweden

Contents

THE SETTING

Chapter 1 The Choice of Policy   3
              The Context of Choice   3
              The Content of Choice   6
              The Profiles of Choice   17

      2 **A Framework for Comparing Policy Choices**   22
              The Choice of Problem and Concepts   22
              The Theoretical Context of Policy Choice   25
              The Political Context of Change   32
              Comparative Policy Analysis   35

THE CHOICE

      3 **Practicability or Principles?**   39
              Possible Policy Alternatives   39
              Sweden Argues the Practicable   41
              The United States Argues the Desirable   51
              Summary   60

      4 **Allocation of Policy Authority and Responsibility**   62
              Concepts and Arguments   62
              Sweden Argues for Central Administrative
                  Authority   65
              The United States Argues for Increased Federal
                  Assistance   72
              Summary   80

      5 **Public Participation in the Implementation of Policy**   83
              Concepts and Arguments   83
              Sweden Argues for Limits to Participation   85
              The United States Provides for More
                  Participation   92
              Summary   99

xii *Contents*

6 Explaining the Differences in Policy Choice   102
    Physical and Environmental Considerations   102
    Technological Considerations   108
    Socioeconomic Considerations   110
    Political Considerations   116
    The Hare and the Tortoise—Two Patterns of Choice   125

THE CHANGE

7 The Hare is Getting Tired   131
    Retreating from Statutory Deadlines   131
    Auto Emission Deadlines Suspended   132
    Adjusting Compliance Deadlines for Stationary Sources   142
    A Strategic Retreat on Objectives   150

8 The Tortoise Keeps Moving   159
    Moving Forward Ever So Slowly   159
    Establishing New Guidelines   160
    Protecting Sweden from Sulfur in Fuel Oils   162
    Increasing Control of Auto Emissions   169
    Approaching Objectives Step by Step   176

9 The Hare or the Tortoise?   181
    Polity and Politics   181
    Politics and Policy   185
    Policy and Impacts   188
    Conclusion   195

Notes   197

Bibliography   225

Index   233

THE SETTING

CHAPTER 1

The Choice of Policy

The Context of Choice

On December 5, 1969, a few men met at the Volvo foundries in Skövde, an air-pollution-ridden town in southwestern Sweden. The meeting was a last effort to solve some differences of opinion between the Swedish Association of Foundries and the Swedish National Environmental Protection Board (NEPB). The foundry association was discontented with the emission guidelines proposed by the NEPB in June 1969. These proposed guidelines were to serve as the basic tool for the governmental agencies and authorities responsible for implementing the new Environment Protection Act of July 1, 1969.

To prepare for swift and smooth implementation of the new act, the NEPB and its predecessor, the National Air Pollution Control Council, had formed (as early as 1966) joint expert panels, which included representatives of the industrial sectors regulated by the act. These panels gathered and processed information concerning the technical and economic feasibility of different control techniques, strategies, and goals. The panels especially tried to assess the economic consequences of different levels of emission guidelines, on industrial sectors. During the latter part of 1969, some intensive bargaining took place over appropriate levels of permissible emissions. Swedish industries managed to achieve more lenient overall guidelines in one-sixth of all cases.

The problems of the Swedish foundries were the most difficult ones to solve. At the December meeting in Skövde, the foundry representatives argued that the emission guidelines proposed by the NEPB in June were incompatible with the criteria spelled out in the Environment Protection Act. According to these criteria, industry was not required to take preventive measures against pollution other than those found "economically and technically feasible." Neither the economy of the foundries nor the technology of air pollution control matched the proposed guidelines. Consequently the foundry representatives argued that if the NEPB wanted foundry cooperation in implementing the act, the guidelines must be substantially modified.

The informal December meeting at Volvo in Skövde resulted in some changes in the guidelines, which essentially met the demands of the foundries. Within one week, the NEPB issued and promulgated the final guidelines for emissions from air-polluting industries. The guidelines expressed the agreement on the practical meaning of the act's criteria of "economic and technical feasibility" that had been reached between the NEPB and the polluting industries in the joint expert panels.

Paralleling these efforts were those of another joint group of governmental and industrial experts. Established within the Ministry of Transport, the joint group on automobile exhaust control reached an agreement in 1968 concerning auto emissions. The joint expert agreement was ratified by the Riksdag and scheduled to become effective on July 1, 1970.

These agreements marked the end of an investigatory, consultative, and legislative process of policymaking that had begun in 1963. At that time, a Royal Commission was established to investigate the problem of nuisance arising from the use of property. As it turned out, the commission found itself working with air and water pollution problems before any strong public opinion had manifested itself on these matters. Its 1966 proposal was heavily directed toward the judicial process as the means for solving disputes emanating from uses of property. However, by the time the proposal reached the Riksdag in early 1969, the cabinet had almost completely reversed it, so that, in effect, the implementation of the new Environment Protection Act would be a matter for bargaining—within certain limits—among the administrative agencies and property users concerned.[1]

To the experts of the joint working panels, the 1968 and 1969 agreements on criteria and guidelines meant embarking on a realistic course of air pollution control. The guidelines made the criteria of "economic and technical feasibility" operational, in such a way that implementation of the Environment Protection Act would be possible within the constraints set by technological development, economic realities, and administrative capability. There was some understanding that, in the long run, public health should be the main criterion for air pollution control measures. However, with only limited knowledge of the relationship between air pollution and public health, such a policy course was unrealistic. Thus the *practicable* rather than the *desirable* determined and defined the new Swedish pollution control policy.

In the same week as the Volvo meeting, a process of changing existing policy was set in motion in the United States. On December 8 and 9, 1969, hearings were held in Washington to review existing legislation. A subcommittee of the House of Representatives found that the Air Quality Act of 1967 (a wonder of legal complexity with very intricate rules for federal, state, and local authorities) had in fact not yet begun to be implemented. More than two years after its enactment, responsible authorities on different government levels were still struggling to establish formulas of interpretation and implementation upon which they could all agree.

The House hearings were a surprising intrusion into an area which for most of the 1960s had been a Senate reserve. Senator Edmund Muskie, candidate for vice-president in 1968 and at the time a presidential hopeful for 1972, had played a key role in the development of air pollution control legislation in 1963 and 1967. Throughout, Muskie had followed an incremental policy strategy and made sure that policy goals were matched by implementative capabilities. His own bill of December 10, 1969, envisaged a continuation of this pattern.

However, when the new Clean Air Act Amendments were passed in December 1970, the issue context had altered dramatically. In effect, the House hearings held one year earlier marked the beginning of a process of radical change in air pollution control policy. The legislative and executive branches of government were under pressure from a dramatically increasing wave of public opinion in favor of new and strong antipollution measures. The growth of opinion was symbolized by Earth Day, April 22, 1970. President Nixon presented his environmental message in February 1970, advocating a program of national emission standards, something that Muskie had opposed in 1967. In the House of Representatives, strong antipollution legislation was deliberated throughout the spring and passed in early June.

All this could be seen as competitive maneuvers designed to take the initiative away from the Democratic hopeful, Muskie. On top of it all, the Senate's antipollution champion came under heavy attack from Ralph Nader who, in May 1970, released reports indicating that Muskie had "sold out" radical antipollution measures in 1967. However, Muskie responded in June by hinting that his Senate subcommittee would recommend tough legislation. When the bill was reported in September, it became clear that it had withstood fervent attacks by the industrial lobby in the foregoing weeks. The health of the people was seen as more important than technical practicability. To protect public health, it was found necessary to establish strong national standards and to state explicitly the deadlines within which the standards should be achieved. However, no hearings had been held to investigate the technological ability of the automobile manufacturers to meet the 1975 deadline for a virtually pollution-free combustion engine.

After the bill had passed the Senate by voice vote, it went through conference deliberations. Notwithstanding strong industrial pressure, the conference reported a bill in mid-December 1970, which, in fact, was essentially identical to the one reported by the Muskie subcommittee three months earlier. The bill passed by voice vote in both the Senate and the House, and was signed into law by President Nixon on December 31, 1970.[2]

To the politicians, the dramatic upsurge in national concern with the environment signaled that support of radical policy initiatives could become an important asset in the upcoming elections of 1970. Likewise, it indicated that careful and realistic steering within the limits of implementation capabilities could become a liability. Hence the rush to maintain legislative initiative, and to establish public health rather than technical and economic feasibility as the main criterion for policy action. For an American congressman, exposed as he is to public scrutiny, the impact of the former as compared to the latter goal was quite evident. The result was a drive for quick adoption of legislative proposals, even at the expense of thorough investigations of implementation capabilities. Under the circumstances, the *desirable* rather than the *practicable* characterized and defined policy. Without much attention to the availability of means, politicians set goals for a policy that differed radically from the one they had adopted only three years earlier.

The Content of Choice

In the comparative description of policy alternatives adopted in the United States and Sweden, three elements of policy content will be discussed: (1) major control approaches and means of implementation; (2) distribution of policy competence and authority among different levels of government as well as means and procedures for enforcement and compliance; and (3) provisions for, and means and channels of, public participation in policy implementation and enforcement. The concept of implementation is broadly conceived; it includes both the procedures and mechanisms for goal achievement provided by legislation and the actual processes of interpreting and carrying out the legal provisions. The second category of power distribution also includes an outline of the formal administrative organization. The third category involves a discussion of sanctions available to enforcement agencies and authorities. It should be noted that the description is concentrated on the main policy legislation and organization at the national level as of 1973–74. Swedish policy is found mainly in the Environment Protection Act of 1969 and in subsequent regulations and ordinances; United States policy choice is contained in the 1970 Clean Air Act Amendments and in subsequent regulations and orders.

Control Strategies

It is evident that the two countries have made different initial choices

TABLE 1. Major Initial Approaches to Air Pollution Control in Sweden and the United States

Approach	Sweden	United States
National ambient air quality standards	No—but two standards for SO_2 and particulate matter have been added	Yes—primary standards to protect public health; secondary standards to protect public welfare
Legally enforceable national emission standards	Yes—for auto emissions	Yes—for new stationary sources; for hazardous pollutants
National emission guidelines (not legally enforceable)	Yes—to complement the franchise system	No
A system of franchise for licensing of individual stationary sources	Yes—the core of Swedish pollution control	No—but state review, prior to construction of the location of new stationary sources
Regulation of the composition of fuels	Yes—especially fuels for heating and energy	Yes—especially fuels for motor vehicles
Governmental grants	Yes—five-year program (1969-74) of subsidies to pollution control investments in existing stationary sources	Yes—for the planning and implementation of state and local pollution control programs
Special air quality control regions	No	Yes

regarding the major approach to stationary source control (see table 1). To a considerable extent, the United States approach involves controlling the quality of ambient air as a whole. A principal consideration has been to promote a uniform approach for all of the states. Another factor has been the desire to prevent detrimental public health effects. Sweden, on the other hand, has opted for the control of each individual source. Even if the 1969 act seems to allow for variations because of different geographic and physical conditions, the subsequent guidelines provide for uniform treatment, at least within each industrial sector. As previously noted, considerations of public health played a somewhat limited role throughout the process of policy choice.

Under the 1970 legislation, United States ambient air quality standards were set at two levels—primary standards to protect public health and secondary standards to protect public welfare. In April 1971, the federal Environmental Protection Agency (EPA) promulgated national air quality standards for six classes of pollutants: sulfur dioxide, particulate matter, carbon monoxide, photochemical oxidants, nitrogen oxides, and hydrocarbons.[3] Prior to setting standards, criteria for major pollutants must be established, based on information concerning the effects of pollutants and the costs of preventing such effects. On the basis of new criteria, one secondary standard for sulfur dioxide was revoked in 1973.[4]

The 1970 amendments require each state to take responsibility for air quality within its boundaries, and to determine how national air quality standards are to be achieved within each air quality control region. The states are required to attain the national air quality standards within three to five years and to maintain them through development of air pollution control strategies. The strategies are outlined in the implementation plans, which each state must submit to the EPA for approval within nine months after the promulgation of national standards. The content of the state implementation plans is spelled out in great detail in guidelines issued by the EPA.[5] By November 1973, fifteen state and district implementation plans had been fully approved by the EPA. Forty other plans were in different stages of EPA assessment.[6]

The United States system for realizing national ambient air standards provides for substantial financial aid to the states to develop implementation plans and specific pollution control programs. Special contractual assistance is also provided for specific planning activities required by the federal government. Grants may cover up to two-thirds of the planning costs and one-half of the operating costs of air quality control programs in single communities. Joint programs may receive higher grants, and interstate air quality control regions may get up to 100 percent in federal support in their first two years of operation.[7] By 1973, the EPA had provided grants to 54 states and districts and to 176 regions and local communities. Federal assistance increased from 32 percent in 1965 to 44 percent in 1973, amounting to $113 million.[8]

The Swedish system of individual source control has been described as "so complicated that it might never have been adopted if we would have had to explain it to foreigners in advance."[9] Stated as simply as possible, the Environment Protection Act of 1969 envisages a licensing system that covers all use of property that may cause environmental disturbance. All who are engaged in, or are planning to engage in, polluting activities must have their plans for construction or alteration of plants, factories, and other installations assessed according to several criteria specified in the act. Using such a criterion as "least detrimental location," the licensing authorities may prescribe any protective measures or limitations for the polluting activity that can be reasonably demanded in view of the criterion of "technical and economic feasibility." What is considered reasonable may differ in individual cases, depending on the specific analysis of all relevant circumstances.[10]

In specifying what can reasonably be demanded, the licensing authorities are helped by the system of national emission guidelines mentioned earlier. The 1969–70 guidelines—which were modified in 1973—contain recommendations as to the maximum allowable emissions of certain types of pollutants from specified classes of stationary sources. The guidelines are only recommendations. They are legally binding only if they form part of the conditions laid down in the license itself.[11]

Under the ordinance of 1969, sixty-eight types of plants and activities are covered by pollution control regulations. Thirty-eight of these have the option of applying either to the quasi-judicial Franchise Board for Environment Protection (FBEP) for a *permit* to "carry on an activity endangering the environment," or to the NEPB for an *exemption* "from the duty of applying for a permit." For twenty-five other types, it is sufficient to give advance notice to the regional county administrations.[12] Both the permit and the exemption determine the conditions, limitations, and protective measures to be observed and taken by the polluter. The difference is that while the permit is legally binding and may secure the polluter against further pollution control demands for ten years, the exemption is valid only until further notice is given by the NEPB.[13] It should also be mentioned that for certain classes of heavily polluting industries, the question of where to locate and how to control pollution is settled at the highest government level—by the cabinet.[14]

Another important element in the major Swedish approach is the system of government grants and subsidies for pollution control measures in existing plants and factories. Under the original 1969–74 program, existing plants could receive up to 25 percent in state subsidy. For reasons of economic recession and seasonal unemployment, extraordinary subsidies of up to 50 or 75 percent were given every winter from 1971 to 1974. The sum total of government subsidies to industrial air pollution control investments for the five-year program thus increased to 260 million Swedish crowns (approximately $58 million). After the end of fiscal year 1973–74, the subsidy program covered only pilot projects.[15]

Legally enforceable national emission standards form part of the United States approach to stationary source control but do not exist in Sweden. By 1973, United States standards had been promulgated for three hazardous pollutants with proved detrimental effects on human health. At the same time, United States national emission standards existed for twelve classes of new emission sources.[16] These source emission standards differ from the Swedish national emission guidelines in being legally enforceable. They are similar to the Swedish guidelines in that they are determined with a view to what is economically feasible and technically practicable with regard to pollution control equipment.[17]

Both countries have adopted standards for emissions from mobile sources. Swedish standards concern vehicles only, while United States standards also include emissions from airplanes. By far, the United States regulations of auto exhaust emissions constitute the most remarkable air pollution control measures contemplated by any government. In demanding a 90 percent reduction of hydrocarbon and carbon monoxide emissions beginning with 1975, and a similar reduction in the emission of nitrogen oxides in 1976, they in effect called for an almost completely pollution-free car by 1975–76. The fate of this initial goal will be dealt with further on; it suffices here to indicate that the first suspension of the statutory deadlines was granted in 1973.[18] Swedish regulations copy the United States requirements for 1973 models and apply to 1976 and later models, thereby making Swedish standards more stringent than those of other West European countries.[19] Regulations of the composition of fuels (common to both countries) especially concern the lead levels in gasoline. In Sweden, programs for a gradual decrease of the sulfur content of fuel oil used for heating purposes form an integral part of the control approach.[20]

Administrative Agencies

Both Sweden and the United States have established special administrative agencies for environmental protection. However, there are some differences with respect to their location and status within the structure of government.

In Sweden, the making of environmental policy is a matter for the Ministry of Agriculture, whose competence spans the environmental spectrum from air pollution control to wildlife management policies (see fig. 1). Implementation of national pollution control policies is concentrated in the NEPB. As we have seen, the FBEP also performs important functions in air pollution control. Both the NEPB and the FBEP are independent agencies. Thus the content of individual agency decisions cannot be determined by the Minister of Agriculture. However, commentaries given in cabinet proposals to the Riksdag and in interpretative ordinances serve as important guiding principles for the day-to-day work of the agencies.

As of 1973, the Air Quality Department within the NEPB had branches for industrial, municipal, and mobile source pollution, as well as for noise abatement. The Administrative Department's Exemption

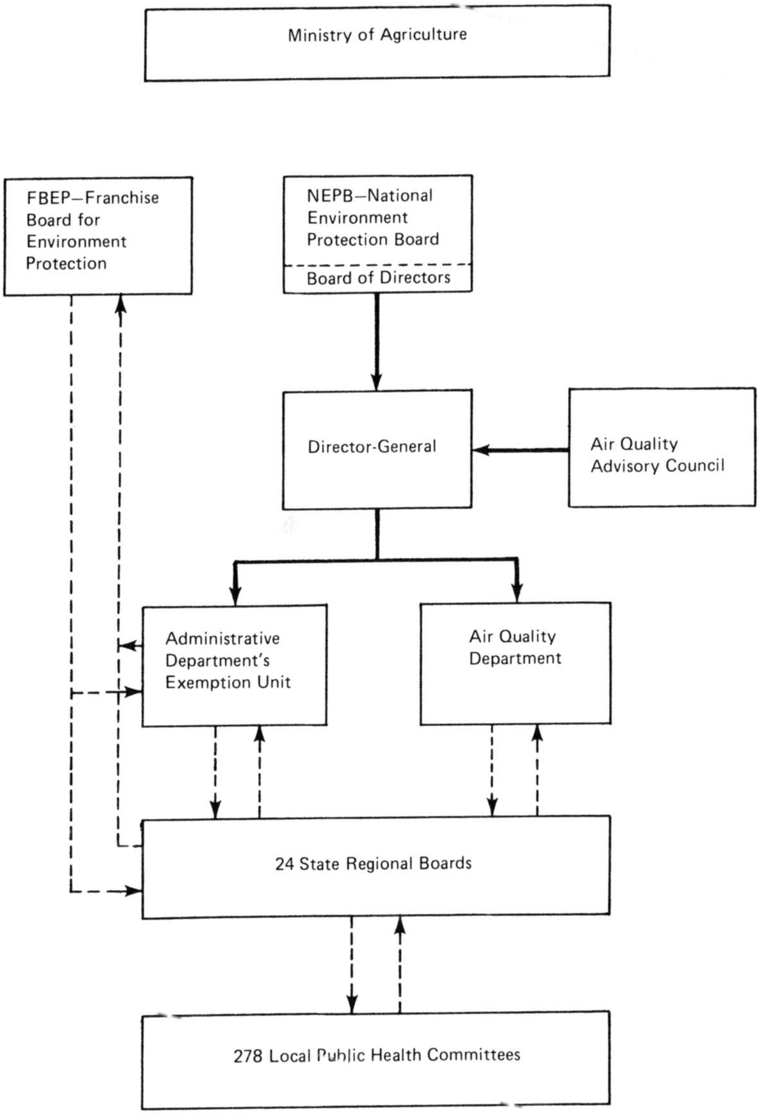

Fig. 1. Organizational framework of Swedish air pollution control, as of September 1973

Unit dealt with issues concerning the licensing of polluting activities. (It should be noted that the NEPB Board of Directors consists of representatives of interests whose activities are regulated in the Environment Protection Act.) Regional pollution problems are dealt with by environment and planning units within the so-called County

Administrations. A better name for these would be State Regional Boards, since they are actually the regional arms of the central national government, and are only to a limited extent responsible to the people living in the regions.[21]

In the United States, there is no cabinet department specifically designated to deal with environmental affairs, although presidential proposals for a Department of Natural Resources were made in the early 1970s.[22] The Environmental Protection Agency is an independent regulatory agency with far-reaching responsibilities. Formally established in 1970 by combining several different agencies and units in the executive branch, the EPA immediately underwent a thorough reorganization and realignment to function as a multidisciplinary environmental protection agency (see fig. 2). Functional and media units were divided, so that the Office of Air Programs was not the only air pollution control unit within the EPA. Also, the offices dealing with research and monitoring, planning, and management, as well as enforcement, performed important functions in the air pollution control policy. Ten regional offices, directly responsible to the EPA administrator, are seen as the "cutting edge" of the agency's efforts. Many of the contacts with the fifty-five State and District Air Pollution Control Agencies are channeled through the EPA regional offices. This would also include all the air quality control regions, many of which are of an interstate character.[23]

With respect to the distribution of competence and authority, one could begin by pointing out some striking similarities in implementation responsibilities (see table 2). In both countries, research and technological development are within central agency competence. Although more than 75 percent of all EPA research efforts are conducted outside the agency, the in-house research team comprised almost 1,900 scientists in 1972, with a research budget of $125 million. EPA also provides technical assistance to the states for development of implementation plans.[24] The Swedish NEPB has the largest environmental research budget in the country, but most of the actual research is contracted to universities and other research institutions.[25] In the United States, the monitoring, so necessary for the setting of ambient air quality standards and for the evaluation of air quality trends, is a federal responsibility. (In practice, however, it is carried out jointly through federal and state programs.)[26] The proposed Swedish Environmental Information System emphasizes central coordination and evaluation of regionally collected data.[27] Another similarity is that standard-setting for mobile sources is a central responsibility in both countries. In the United States EPA is responsible for both setting standards and performance-testing new car models.[28] In Sweden the

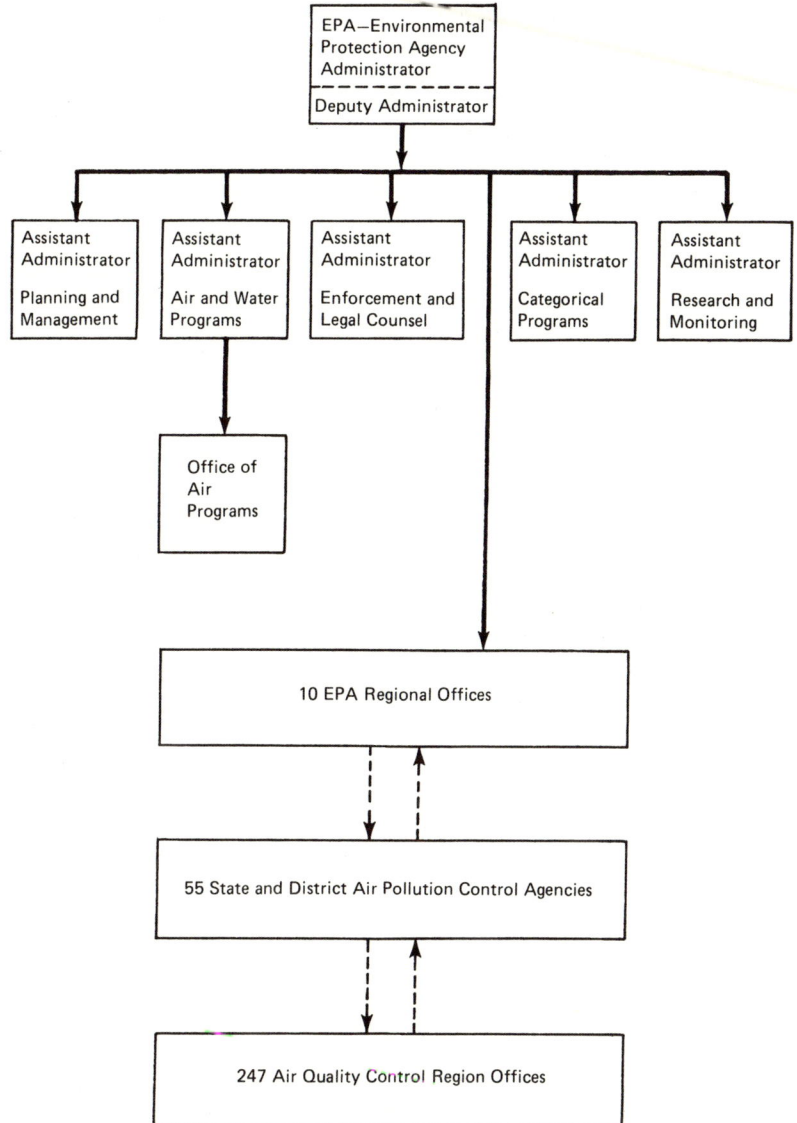

Fig. 2. Organizational framework of U.S. air pollution control, as of November 1972

standard-setting is done by the agency for road and vehicle safety, while performance tests on new models are carried out by the NEPB.[29]

When one looks at the setting of ambient air quality standards and standards for stationary sources, more conspicuous differences become

TABLE 2. **Initial Distribution of Authority and Responsibility for Air Pollution Control in Sweden and the United States**

Policy or Activity	Country and Level of Government	
	Sweden	United States
Research and development	National	Federal
Standard-setting		
Ambient air quality standards	————	Federal
Stationary source emission standards or guidelines		
—new sources	National	Federal
—existing sources	National	State (but federal preemption possible in connection with takeover of implementation plan)
—governmental sources	No special provision	Federal
—hazardous pollutants	————	Federal
Emission standards for mobile sources	National	Federal
Performance testing and control		
Stationary sources	Regional/National	Federal (delegation to states if adequate provisions exist in implementation plan)
Mobile sources		
—new vehicles	National	Federal
—used vehicles	Regional	State/Federal
Monitoring	————	Federal
Licensing	National	Limited state authority
Enforcement	Regional/National	State under implementation plan, federal for federal standards, citizen activity through "citizen suits"

apparent. In the United States, federal action seems to be gradually circumscribing and preempting state and local authority. As has already been mentioned, implementation of federally determined ambient air quality standards is carried out through state implementation plans, subject to federal approval and dependent on federal grants. The EPA has established guidelines for the proper content of state plans. Another important factor is the EPA classification of air quality control regions by severity of pollution and necessity for specific control measures. Still another is the requirement—initiated through court action—that antidegradation clauses be included in state implementation plans. Furthermore, states may be required to control indirect sources and to devise control programs for growth areas. State responsibility may also be preempted through direct EPA action. By July 1974, the EPA had promulgated its own compliance schedules for more than 8,000 facilities, instead of awaiting state action. The EPA may also take over state responsibility for inspection and data collection to control individual polluters' performances and maintain them on the compliance schedule.[30]

In contrast, the Swedish situation from the outset has been characterized by a high degree of centralization of policy responsibility. The process of determining national emission guidelines, where the NEPB made its decisions on the basis of information provided by a highly centralized system of joint governmental-industrial expert panels, has already been discussed, as has the licensing system, which is in principle very centralized, with the FBEP and the NEPB as principal agents. The NEPB has important coordinating functions with respect to supervision and control of polluter compliance, although the actual inspection is carried out by the state regional boards.[31]

Enforcement
This brings us to a discussion of enforcement. In Sweden, enforcement powers are divided between the NEPB and the state regional boards. If a polluting activity is causing "substantial nuisance," the enforcement agencies must first give advice and instruction on suitable action. Certain sanctions may be invoked if action is not taken. If polluters have no permit from the FBEP, the state regional boards can issue injunctions to prescribe protective measures, prohibit the activity, and/or determine penalties. When there is a violation of permit conditions, regional boards may order polluters to rectify the matter at their own expense. There are also provisions for fines, and even imprisonment, in cases of permit violation. At the central level, the NEPB may ask the FBEP to prohibit a polluting activity or to prescribe certain protective measures. When a public interest is affected, the NEPB can request police assistance in exercising its supervisory

powers.³² The 1969 act entitles those directly affected by pollution to file compensation suits in the courts, but contains no provision for citizen activity in policy enforcement. In practice, the enforcement powers have been seldom used. Most cases are settled in a bargaining process involving "advice and instructions on suitable action."³³

The United States enforcement program has three characteristics: (1) the possibility of federal preemption of state powers, (2) greater reliance on the court system, and (3) the incorporation of citizens in the process. Normally, states with implementation plans deemed adequate by the EPA carry out the bulk of enforcement activities. If states are inactive or do not perform their duties adequately, the EPA may notify the states and take over. The agency may enforce the states' own implementation plans. It also has special emergency powers that come into force whenever there is an "imminent danger" to public health. Enforcement powers of the EPA administrator involve, for example, bringing civil action in federal courts against violators of federally approved implementation plans, federal standards, or abatement orders. Sanctions also include fines and imprisonment.³⁴

The EPA has engaged "directly and forcefully, in a full range of enforcement actions." During the first ten months of 1973, agency enforcement actions increased by 100 percent compared to the previous year. Federal enforcement is the rule with regard to mobile sources, as in the instance where the power to recall auto models has been applied.³⁵

Some EPA enforcement activities in 1973 were prompted by citizen complaint. Under Sec. 304 of the 1970 Clean Air Act Amendments, citizens have the right to sue in federal district courts to enjoin violations of the act or the plans, standards, and orders issued under the act. Citizens may also sue the EPA administrator to require him to perform the mandatory functions given to him through the act.³⁶

Public Participation

The most conspicuous differences in policy choice are found when one comes to the issue of public participation in the policy process. Except for the general Swedish rule of public access to agency files, the channels for participation are either indirect or few. Special interest organizations may influence the NEPB decisions through their representatives on its board of directors. As already described, the preparatory work on emission guidelines was a matter for the NEPB and the industrial sectors concerned. There were no public hearings nor was there a period of consultation with a broader spectrum of interests. The permit procedures of the FBEP provide for advance public notice, site inspections, hearings, and appeals, but only for "affected parties" and not for citizens in general. The exemption procedures of the NEPB provide for site inspections and hearings with concerned parties only if that is found necessary for the investigation. There is no appeal from exemptions issued by the board.[37]

The United States situation is substantially different. Admittedly, provision for open EPA and state files as a prerequisite for EPA promulgation of state implementation plans bears some resemblance to the Swedish situation. But more important are the provisions for public hearings and for public participation in policy enforcement. State implementation plans are subject to the scrutiny of public hearings. The EPA holds public hearings on matters of auto emission control. These hearings are much more public in character than the ones provided for in Swedish policy. Furthermore, United States citizens can engage in policy enforcement by filing suits in federal courts against polluters violating the act. Citizens may also sue the EPA administrator if he fails to perform his mandatory functions. The development of class action suits in the area of environmental policy also seems to have broadened the possibilities for the general public to engage in the policy process. By 1973, a considerable number of citizen suits brought under the 1970 act were pending in the courts.[38]

The Profiles of Choice

The most important results of this configurative description of policy choices can be summarized in the following way:

- The Swedish process of choice had a long duration, with experts telling the politicians what could possibly be done and with politicians subject to little, if any, public pressure. On the other hand, the United States process was short and intensive, with the politicians telling the experts and special interests what should be done. The politicians were under considerable pressure from the public.
- Policy choices were mainly regulatory in character, with the distributive elements serving as complements to the major approaches. Only in Sweden was a distributive element chosen to affect the behavior of nongovernment groups.
- Chosen regulatory policies, especially the ones first selected, showed several differences with regard to the scope, area, and intensity of action. Subsequent choices or secondary approaches had much more in common.
- Selected policy alternatives differed with respect to the location of policy authority and responsibility among government levels.
- Great differences existed with regard to the arrangements for

public participation in the policy process, both in the number of open policy stages and channels as well as in the scope of citizens' rights.

Regulatory air pollution control policies may vary from the air resource management (ARM) approach to the individual source control (ISC) approach (see fig. 3). The choices of initial approach differ between the two countries. Sweden is at the ISC end of the spectrum. Its approach covers nearly every polluting activity of any duration, and almost every polluting industrial activity is now subject to some kind of individual source control. What is remarkable is the almost total lack of an ARM approach. For the first five years of the pollution control program, ambient air quality standards existed only for sulfur dioxide. The trend, however, was toward adopting the ARM approach. "There is a need for ambient air quality standards to assess what activities are most necessary from the *environmental* point of view. There are continuing efforts to define ambient standards for particulate matter, and we are reexamining present standards for sulfur dioxide."[39]

The United States approach and trends are almost completely the reverse. Initially largely geared toward the ARM approach, United States policy shifted somewhat with the 1970 Clean Air Amendments and subsequent regulations. Deliberate efforts have been made to combine the two. The United States national ambient air quality standards are directly linked to several types of individual source control. These range from the state standards and compliance schedules for existing sources to the EPA standards for new emission sources. The EPA has taken part in the regulation of thousands of existing sources formally under state control, to ensure that the behavior of individual polluters is consistent with ambient air quality standards.

An important corollary to the difference in initial approach should be noted. To a considerable extent, the approaches govern the nature and organization of data collection and evaluation in the policy process. The United States with its ARM approach has monitoring systems giving continuous information about the trends and developments in the air mass above the nation as a whole.[40] In contrast, Sweden with its ISC approach did not contemplate such a system until the mid-1970s. The difference in the types of data and knowledge is amply illustrated by the problems confronting those trying to evaluate the impact of the 1969–74 subsidy program. For Sweden, only the reductions in emissions from each individual source could be calculated. No information concerning the trends in total air pollution could be deduced from available data.[41]

However, our configurative policy description also points to a certain trend toward centralization of policy powers and respon-

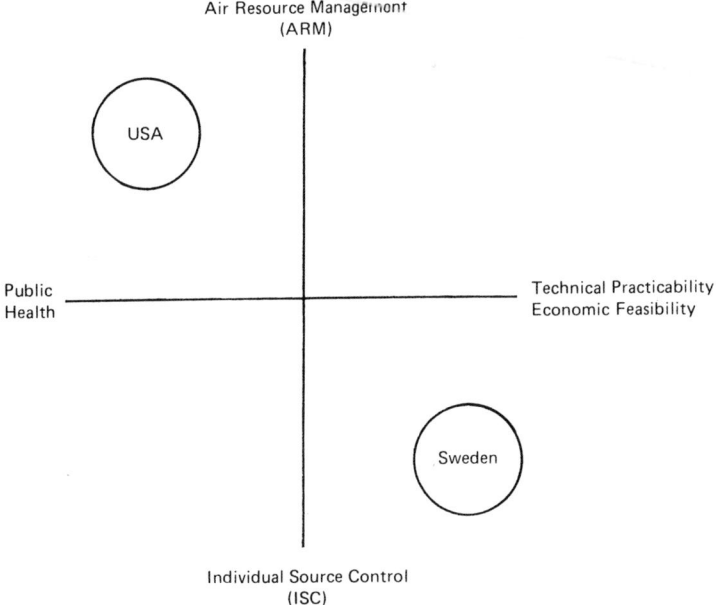

Fig. 3. Main policy principles and major initial approaches to air pollution control in Sweden and the United States

sibilities. Judging from formal statements and recommendations in existing laws and regulations, United States policy responsibility is shared between the EPA and the states, with important functions also allocated to the local authorities. However, examination of United States policy has hinted at the existence—at least for the first few years after 1970—of a propensity to use preemption and expansion of the federal role whenever it is deemed necessary for the achievement of important policy goals. In Sweden, the dominance of authorities at the central government level is complete. Nothing resembling decentralization of authority had manifested itself by 1977.

Regulatory approaches and power distributions are but parts of a larger pattern or process in which ultimate success is measured by changes in the behavior of specified target groups. Of central importance in that process is the nature and availability of means of enforcement and implementation. This also applies to the character and number of channels available for policy enforcement and implementation. The availability and actual use of the courts and other judicial means to obtain preferred behavioral changes implies a policy quite distinct from one where change is attempted through administrative advice and persuasion. This is even more so if the latter approach is linked to a system of economic incentives to change target group behavior.

Considerable differences exist between the two countries. In Sweden, it is fair to say that the mainly *administrative* process of enforcement and implementation is characterized by advice, persuasion, and voluntary compliance. In the United States, even if it is true that the EPA places some emphasis on voluntary compliance, the federal "fair-but-firm" approach implies the use of *judicial* channels, such as the federal courts, whenever such measures are deemed necessary to obtain compliance with policy regulations.[42] This difference between an administrative/consensual and a judicial/adversary process in enforcement and implementation seems to be firmly established between the two countries (see fig. 4).

These differences seem to be related to differences in the openness of the process of policy implementation. Whether one looks at who has access or where points and channels of access are located, the differences are quite conspicuous. In Sweden, there is a system of opposite numbers between the NEPB and the many factories and plants polluting the air, but very few avenues of participation are open to the average citizen breathing the polluted air. As we have seen, the type of data and knowledge assembled under the Swedish system does not particularly enhance the ability of members of the general public to protect their

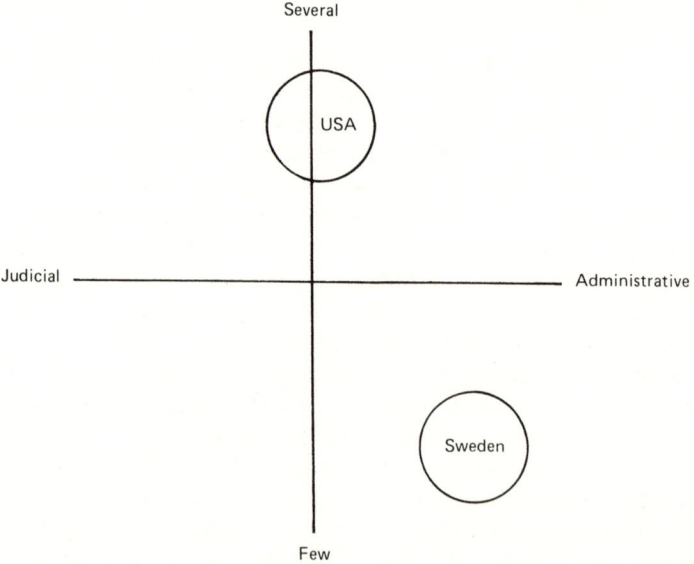

Fig. 4. Channels for implementation and enforcement and number of possibilities for public participation in Swedish and U.S. air pollution control policies

interests as citizens. Both the number of participants and the channels and means of participation open to them are considerably higher in the United States. It must also be noted that environmental legislation, such as the National Environmental Policy Act of 1969 and the 1970 Clean Air Act Amendments, has been central and instrumental in the process of opening up the American policy process. Members of the general public have the right as citizens to participate in policy implementation.

Thus, wide differences exist in the means originally chosen to provide some sort of solution to a problem that, at least at face value, might be interpreted as common to both countries. The profiles of these choices seem to imply that the broader characteristics of each country's system of politics and policymaking, as they are judged and interpreted by actors involved in the process of choice, are of considerable importance in determining the selection of policy. Of course, one must not discount the importance of other factors not immediately considered as political. To assess the relative importance of political factors, there is need for some tools with which to analyze the considerations of central political actors actively engaged in choosing among policy alternatives.

CHAPTER 2

A Framework for Comparing Policy Choices

The Choice of Problem and Concepts

Air pollution knows no boundaries. Metropolitan areas and industrial towns in developed countries are all blighted by pollution from motor vehicles, factories, and energy-processing plants. The political problem of controlling air pollution confronts all urbanized and industrialized countries.

For both scientific and practical reasons, this commonality of pollution problems makes pollution control an interesting field for comparative policy analysis. From the scientific point of view, comparative political analysis has a primary commitment to collecting, ordering, and broadening our theoretical and empirical knowledge of politics. The simultaneous appearance of environmental quality as a political problem in industrialized and urbanized countries suggests that such knowledge can be acquired through comparative analysis of pollution control policies.

From the practical point of view, a comparative analysis of selected and pursued policy alternatives may provide valuable insight into the reasons why certain alternatives are more successful than others. But however timely and practically relevant such comparative studies might be, they may still be somewhat premature because of the relatively short lifespan of most pollution control policies. The practical experiences of such policies are still somewhat limited—and perhaps inconclusive. Asking *which* policy alternatives have been adopted in Sweden and the United States takes us no further than to a configurative description of policy events and contents. Such descriptions do reveal regularities, similarities in governmental responses to common problems. However, it is much more challenging to ask *why* particular policy alternatives were adopted and implemented. The desire for explanation rather than mere description compels us to carry our comparative policy analysis further to find out what factors were decisive in forming the policy alternatives.

As will be discussed, efforts to explain policy content may proceed along two fundamentally different lines, one causal and the other intentional. The main thrust of causal policy analysis is to find statistical associations and correlations among policy content and surrounding social, economic, and cultural determinants or forces that are strong enough to explain policy content as caused by these surrounding factors. To calculate these correlations, the causal policy analyst is often forced to regard determinants and policy content as sets of numerical indicators. Consequently, policy as a choice among alternative sets of policy contents is seldom accounted for. Comparisons start with existing policies and give no account of the process of choice that led to an alternative being selected as governmental policy.[1]

In contrast, intentional policy analysis concentrates on studying policy content as a result of a conscious process of assessing, evaluating, and selecting a particular course of action among numerous possible alternatives. Instead of numerical indicators, this line of analysis focuses on policymakers as goal-oriented problem solvers, capable of evaluating different alternatives in terms of their contributions to the achievement of policy goals. Instead of correlations, intentional analysis concentrates on the perceptions of reality, the calculation of outcomes, and the motivations for choice predominant among key policymakers.[2]

It is the latter perspective that guides this comparative study of clean air policy choice and change in Sweden and the United States. Our general research problem can be formulated as follows: "Why have policymakers in two highly industrialized nations—Sweden and the United States—selected and pursued different policy alternatives to cope with common and technically similar problems of air pollution, and what have been the fates of these policy alternatives?" In particular, the study will focus upon such questions as:

- What similarities or differences can be found in key policymakers' perceptions and evaluations of the physical-environmental, socioeconomic, and political contexts surrounding the choice of clean air policy alternatives, and how are these similarities or differences reflected in their recommendations for clean air policy action?

- In particular, are there any similarities or differences in key policymakers' assessments of the importance of political institutions and structures in the choice of clean air policy?

- What similarities or differences can be found in key policymakers' preferences regarding the goals to be pursued in clean air policy, and how are such similarities or differences reflected in recommendations for policy choice?

- How do key policymakers perceive the impacts of selected and pursued policy alternatives and of changes in the physical-environmental, socioeconomic, and political context of clean air policy, and what are their recommendations for policy change?

Before discussing further the implications for comparative policy analysis of this intentional perspective, some discussion of certain concepts is in order. This is a study of environmental policy; unfortunately, the concept of environment has many cultural as well as scientific connotations. One political scientist has concluded that "on neither technical, economic, or political criteria is it possible to make entirely satisfactory definitions because of the interpenetration of environmental policy with other fields of policy analysis."[3] I shall use the following definitions:

Environmental policy is a course of action chosen and pursued by government to cope with the social problems of physical environmental quality. The physical environment includes the elements of air, water, and soil with their ingredients and properties. Here, it is not viewed as including the inner parts of such man-made surroundings as houses and factories, although it is recognized that activities taking place there may have to be subjected to government action to achieve a desired level of physical environmental quality. Clean air policy is thus preoccupied with air quality in the physical environment so defined. The social problems of environmental quality—here, air quality—connote environmental impacts of human and societal activities that are perceived to be harmful to environmental quality as well as to public health and to the quality of human life.

Public policy is used here to connote a selected and actually pursued course of government action within a physical-environmental, socioeconomic, and political context that provides the specific opportunities and obstacles that the policy is supposed to overcome and/or utilize to reach a goal or realize an objective.[4] Policy is viewed as a problem-solving activity, involving goal-conscious actors. With such a view, "the interesting questions of policy analysis become those of . . . public choice, of how those in positions of authority, as well as those who advocate alternatives and otherwise engage in the political argument, go about defining an appropriate response to a given set of circumstances, and how appropriate their response turns out to be."[5]

The concept of *policy choice* refers to the selection and adoption of one policy from among many available alternative courses of government action. A crucial assumption underlying the intentional line of policy studies is that alternatives are not considered equally feasible or desirable. Only a finite number of alternatives are perceived as possible or feasible, and an even smaller number as *desirable* solutions to a given social problem.

The selected and pursued *policy alternative* is defined as the particular allocation of resources designed to achieve a stated public purpose. Such an alternative possesses at least the following five characteristics: (1) a stated intent, that is, an authoritative expression of intentions with regard to a preferred conduct or state of affairs; (2) a programmed resource commitment, as reflected in budgets, plans, and organizational arrangements; (3) actual resource flows—resources must be used, programs executed, and organizational duties performed; (4) provisions for inducements or sanctions—positive and/or negative means of establishing the preferred conduct; and (5) performance and effectiveness; in other words, there must be a discernible impact of the policy upon the particular social problem.[6]

Clean air policies vary considerably with respect to all these elements, despite some fundamental and general characteristics of the policy problem. From a technical point of view—focusing on the relative concentration and diffusion of air pollution control costs and benefits—clean air policy is characterized by concentrated costs (polluters) and diffused benefits (all breathing human beings). This is so because of air quality's character of collective good. From a structural and political point of view—focusing on the character of coercion and inducement and the probability of their use—clean air policies would, as a rule, be regulatory because the policy problem is one of correcting or achieving a particular behavior or habit of specific individuals.[7]

The intentional line of policy analysis recognizes that each policy choice is surrounded by a unique set of circumstances, and that variations in policymakers' assessment and evaluation of contexts and alternatives make for wide variations in policy, regardless of the general theoretical similarities of the policy problem. A framework to guide such intentional policy analysis is what we turn to next.

The Theoretical Context of Policy Choice

To view policymakers as problem solvers has several implications for comparative policy analysis. Politics and policymaking are viewed as a process of conscious assessment of possibilities and constraints offered by a specific physical-environmental, socioeconomic, cultural, and political context. Furthermore, the assessment of policy alternatives is seen as being made consistent with a policymaker's value structure or order of preferences. This implies that the answer to why different policy alternatives have been chosen to solve the same policy problem can be found by analyzing the actions, motivations, and arguments of the key policymakers in each country.

But this perspective raises important problems in comparative policy analysis. First, one must find a framework that makes it possible to relate variations in selected policy alternatives to variations in the given set of circumstances surrounding the policy choice. Second, the framework must be linked to the view of policymakers as problem solvers. Third, the framework should make possible an ordering of the elements entering the actor's choice of policy alternative. Fourth, the framework must make possible a comparison of the perceptions held, the motivations and arguments presented, and the choices made by key policymakers in different countries.

In relation to the first problem, it seems preferable to adopt the view that "international diffusion of policy technique" is the normal situation, at least for industrially and technologically developed nations. New control techniques and new means of inducement and coercion become rapidly spread among states, but there are obstacles and blockages to their adoption in individual nations. New techniques are available simultaneously in many nations, but there are leads and lags in their adoption as national policies.[8] This suggests that differences in policy may not be random but subject to variations in accordance with the strength and configuration of the given set of circumstances surrounding the process of choice.

Such a view does not necessarily include policymakers as problem solvers. It may as well put the researcher in the camp of analytical-correlational policy studies mentioned above, where policies and their contents are studied as if the perceptions, arguments, and choices of policymakers make little or no difference. The correlations between indicators of the different aspects of the context and of the policy content are calculated with little reference to policymakers and their assessment of feasible and desirable policy alternatives.[9] Therefore, our analytical framework must also satisfy the other three criteria, for we suspect that politics and politicians do indeed matter in the process of designing and adopting courses of government action.

Thus, underlying the view of policymaking as a process of choice and a problem-solving activity is the assumption that those who make policy are rational actors. Our assumption is not as far-reaching as those of the formal, deductive models of political economy and public choice, of game theory or coalition theory, which attempt to define the conditions of political rationality under postulated abstract conditions. All we assume is that policymakers, in interpreting the demands for action and the supply of policy alternatives, are capable of assessing the costs and benefits of different alternatives, of identifying the outcomes of different alternatives, and of assessing—however tentatively—the probabilities of these outcomes. We also assume that policymakers are capable of ranking outcomes in some order of preference, and of relating them to their goals and objectives. We do not postulate or assume that policymakers have a comprehensive and full knowledge of all possible alternatives or all imaginable outcomes. But it is assumed that they will generally avoid alternatives perceived to have negative consequences, such as increased costs and decreased benefits.

This is the assumption of implicit rationality that underlies our comparative analysis.[10] Far from pushing politics aside, it helps to place politics and policymakers at the center of our comparative study. It is their perceptions, their calculations and assessments, and their recommendations that form the fluid from which the policy alternatives are finally distilled.

The assumption of implicit rationality makes it possible for us to analyze the process of policy choice as a calculus made by political actors (see fig. 5). This calculus can be divided into four distinct subcalculi, which ought to be kept apart analytically: (1) the substance calculus, (2) the preference calculus, (3) the strategic calculus, and (4) the final balancing, or trade-off calculus.[11]

The *substance calculus* involves assessments of what alternatives are available. It furthermore involves an assessment of the outcomes associated with each of the available alternatives, as well as an assessment of the probabilities of each of these outcomes. The *preference calculus* involves an ascription of values to each of the probable outcomes, at least in the sense that they are ranked in some order of preference by the policymakers. The *strategic calculus* consists of assessments of the probable behavior of other political actors—be they parties, individual politicians, voters, or pressure groups—if a certain alternative is recommended as *the* policy. (For example, if party A prefers one line of action and recommends its adoption as government policy, this might diminish its electoral support or its chances of entering a future coalition government.) The *balancing* or *trade-off calculus* consists of an integration and final intercalibration

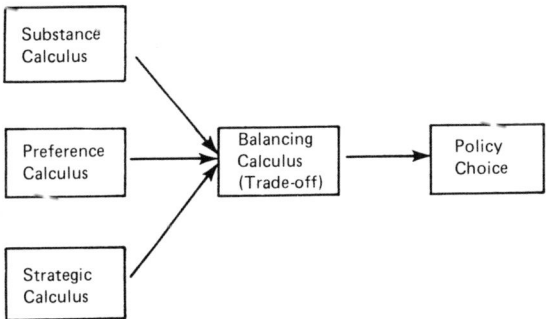

Fig. 5. A general model of the main elements in any process of policy choice

among the first three elements of choice to reach a final choice with the most favorable overall consequences.

So far, we have discussed the problem of policy choice in very general terms, which could be applied to any policy choice. But we also want to outline the specific considerations and calculations of clean air policy choices to make complete our analytic framework (see fig. 6). It is clear that physical geographic and climatic conditions must play a central role. Countries vary in size, topography, and climate, and these variations may present clean air policymakers with very different pollution control problems in different countries. However, with the exception of such natural air pollution problems as those caused by volcanic outbreaks or dust storms, most air pollution is man-made and

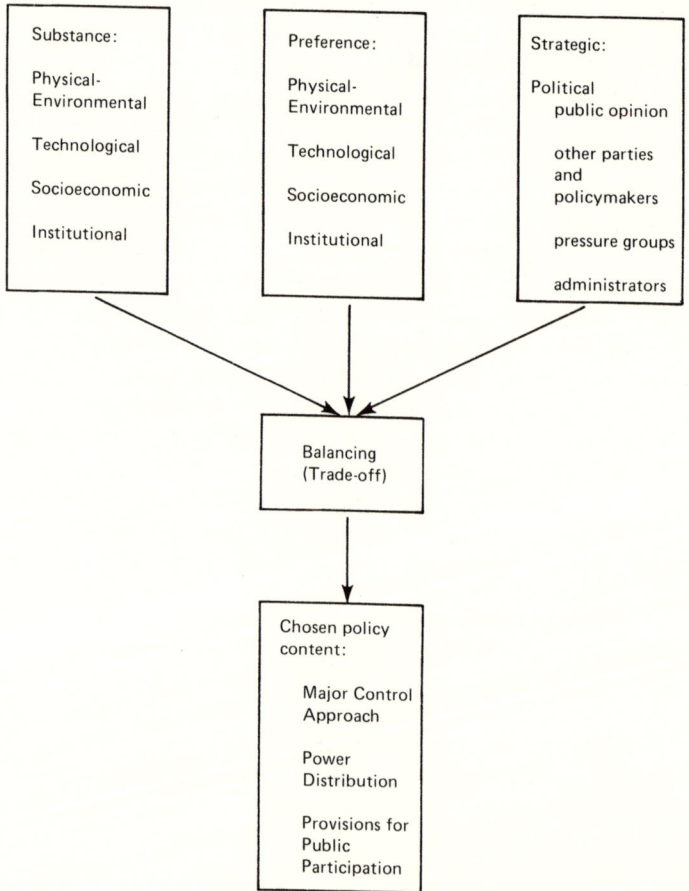

Fig. 6. A model of the specific elements in the process of selecting clean air policy alternatives

closely related to socioeconomic conditions and activities, such as industrialization, urbanization, and transportation. Technically similar pollution control alternatives may have different socioeconomic consequences because of different socioeconomic structures in the countries; that is something clean air policymakers must assess and evaluate. They are thus faced with the necessity to consider pollution control technologies and their availability, consequences, and acceptability. The choice of a governmental course of action to achieve clean air is an inherently political activity. Clean air policymakers must assess policy alternatives in terms of consistency with, or consequences for, existing formal *institutions* of government, the *substance* of earlier policies, and so forth. But they must also assess policy alternatives in terms of *strategic* political consequences—the behavior of other actors, the possibilities of coalitions, etc.—and decide whether they find these consequences preferable or not.

This is the analytic framework I used for my comparative analysis of why Sweden and the United States made different clean air policy choices in 1969–70. I wanted to find out what considerations were most important to key policymakers as they set out to make their choices. Were they environmental, socioeconomic, or technological? Or were political considerations decisive in forming the policy choice? If so, what kind of political considerations? To answer these questions, the analysis will focus on key policy documents in the two countries, to order and interpret the arguments and motivations given by key policymakers for or against different policy alternatives.[12]

Without going so far as to formulate specific hypotheses, let us try to illustrate some of the explanations our comparative analysis might suggest, by putting ourselves into the position of "surrogate policymakers" faced with the problem of controlling air pollution.[13] Choosing a policy for stationary source control immediately invites questions concerning the physical-environmental variations in the country. If these do not contribute to any significant variations in pollution among regions with similar business structure, a single national approach might be contemplated. If, on the other hand, physical-environmental conditions result in different regional air pollution problems within the same industrial sector, a more varied, regional approach might be warranted.

Whether we contemplate a national or a regional approach to stationary source control, we have to consider the socioeconomic and political consequences of each of these alternatives. While a single national approach leads to competitive justice for all firms within the same industrial sector regardless of geographic location, it might indeed kill a whole region if that region is socioeconomically dependent on a single air polluting industry. Furthermore, a single national approach is easily adopted if there are no regional bodies with constitutional jurisdiction over socioeconomic activities that create air pollution problems. Of course, if we prefer centralizing political and administrative power, we can adopt a national approach despite the existence of such constitutional hindrances, even if that would mean changing the distribution of power within existing clean air policies.[14]

As surrogate policymakers, we will soon be aware of the fact that air pollution has serious effects on public health. However, we will also be aware of the fact that pollution control technologies are expensive, and that demanding their use will have far-reaching socioeconomic consequences. If our main policy preference is to protect public health, we may lean toward adopting an alternative that "pushes" the development and introduction of new and more efficient pollution control technology. But if we are more concerned with the socioeconomic consequences of pollution control, we may lean toward an alternative that adjusts to new technology as it develops. We may even contemplate subsidizing the installation of pollution control technology to avoid disruptive socioeconomic consequences.

Our approach to technology will be highly dependent on our assessment of the political context of choice. As environmental ministers in a parliamentary democracy, we must realize that we will be held immediately responsible for any disruptive socioeconomic consequences of expensive control technologies. If, furthermore, we are elected as party representatives from multi-member constituencies, we might be somewhat shielded from fluctuations in environmental opinion. As a result, we may settle for an adjustment strategy as a safe alternative in terms of foreseeable consequences. Since our opponents may very well enter a future government, they also know they will be held immediately responsible for potentially disruptive alternatives. This will keep them from proposing radical policy departures, however popular these may be in environmental opinion.

If, on the other hand, we are legislators in a balance-of-power system, we know we will not be held immediately responsible for negative policy consequences; they can be blamed on bad executive implementation of legislative intent. Provided that we are personally elected from single-member constituencies, and also provided that there is strong public opinion in favor of radical policies, we might be ready to gamble by choosing policy alternatives involving a technology push with uncertain socioeconomic consequences.[15]

As we move to the field of mobile source control, we find that many of the constraints and possibilities just mentioned are present. However, there is one important difference. Moving sources do just that; they move about, across jurisdictional borders. However well founded the arguments for a regional approach may be, there are strong socioeconomic incentives for a unified national approach to mobile source pollution control.

But the process of choice does not end there. Pollution control requires technological development. The devices are expensive to develop, buy, and maintain, and the socioeconomic consequences must ultimately be borne by someone. As surrogate policymakers in a small country, with a relatively limited market for mobile sources, we might find the consequences of a technological push for automobile owners, the national car industry, and transportation policies too great to warrant its adoption. As policymakers in a large country with the world's largest auto industry exercising its political clout and with an existing technology adjustment policy, we might come to the same conclusion.

However, the political context of choice may be such as to narrow down the number of available alternatives. A strong public opinion may be pressing for the adoption of radical policy alternatives. As surrogate policymakers, many of us up for reelection soon, we may have little choice but to propose radical policy alternatives to gain votes and get reelected. The more institutionally competitive the situation of policy choice, the more important will such strategic calculations be in most policymakers' minds.

Again and again, we come back to the importance of public opinion in the considerations of key policymakers. The emergence of environmentalist groups, often with very sophisticated ways of communicating their demands to policymakers, has occurred at the same time that industrialized countries have adopted new environmental policies. One might be tempted to conclude that activities of environmental pressure groups have motivated politicians to introduce more stringent air pollution control strategies. On the other hand, environmental opinion may have been nothing more than a diffuse, albeit widespread, discontent with pollution, not amounting to a strong, coherent pressure group activity for adoption of specific policy alternatives. Policymakers may thus have had a wider latitude of policy choice than is generally assumed.

This very tentative outline of some of the problems facing policymakers charged with selecting a government clean air policy has attempted to indicate the types of physical-environmental, socio-economic, and technological considerations entering the process of choice. I hope it has also indicated the efforts that were made to find out how, and to what extent, perceptions of the political context and calculations about the strategic political consequences of alternatives influence key policymakers' recommendations for action. The closest we have come to formulating some sort of hypothesis is to state that the more institutionally competitive the political context, the more important strategic calculations about the political consequences will be to most key policymakers, and the more inclined these politicians will be to propose and recommend radical policy alternatives. Cheered by the audience, the hare will dash away with great leaps, while the shielded tortoise will move forward slowly but steadfastly. Does this also mean that the hare will get tired or change speed and direction more easily than the tortoise?

The Political Context of Change

A comparative study of policy development that stops at the moment of choice is literally a study of no consequence. A policy choice acquires meaning mainly through the actual use of the means provided for the implementation of its intentions. One could argue that ultimately the tangible results of a policy are more important than how the policy came about, provided, of course, that the implementation does not violate general norms concerning proper governmental conduct.

Studies of the tangible changes resulting from policy implementation are generally referred to as *impact studies*. One looks for discernible changes in those physical-environmental, socioeconomic, technological, or political conditions intended to be changed through the implementation of the adopted policy alternative. The chosen policy alternative is followed forward in time to trace the various changes effected by its implementation. The perspective is causal; the assumption is that the policy causes the changes observed. Of course, there may be other causal factors at work besides the policy, but the main interest is focused on policy as the independent variable.[16]

However, this is not the only perspective applicable to a study of policy and change. One might also be interested in following the development of the original policy choice and studying the changes being made in policy objectives and the means provided for implementing these objectives. It is this perspective of policy and change that will guide the third part of this study.

With such a perspective of policy and change, there is no need to shift from an intentional to a causal mode of policy analysis. Such an approach to change is fully consistent with an analysis that focuses on the perceptions, assessments, motivations, and recommendations made and given by key policymakers. It is their perceptions of the impact of chosen and implemented policy alternative, their assessment of the impact on policy of changing conditions, and their views and recommendations concerning the relationship between clean air policy and other social issues that are of central concern. We will analyze how key policymakers answer such questions as "Is this policy getting anywhere? Is it consistent with the goals and objectives in other policy areas? Has the policy context changed, either as a result of the policy or as a consequence of other developments, in a way that warrants a change in the existing clean air policy?"

The first question is concerned with the problem of consistency between policy objectives and policy means. Are the resources set aside sufficient to achieve the goals? Are they used efficiently? Are the goals really achievable with present and programmed resource commitments? If policymakers suboptimized in the first place, that is, if they set goals less ambitious than warranted by available and foreseeable resources, they may be tempted to adjust the objectives upward. On the other hand, if policymakers were gambling, if they set policy goals beyond capability, they may be tempted to engage in a strategic retreat on objectives, rather than providing the resources necessary.[17]

In clean air policy, this question has to do with whether or not there are technologies available to achieve certain levels of pollution control or air quality. A technology-push approach might soon be found impossible to implement because the development timetable is much longer than expected. Consequently, policymakers may regard a change in objectives as more attractive than the adverse socioeconomic and other consequences that might follow from continuing with the push approach. On the other hand, policymakers who suboptimized in the first place by adopting a technology-adjustment approach may find it possible to adjust the standards of pollution control and air quality upward. In the end, the tortoise may beat the hare, because the hare overrates his capacity.

As the example indicates, policymakers may find it necessary to change policy not only because of tensions inherent in the policy itself but also because of what they perceive as unintended or unacceptable consequences for other policies and areas of social concern. Air pollution control may have consequences in the areas of energy supply and transportation policy that may not be consistent with social and political preferences for welfare and economic development. If such consequences coincide with dramatic events, as was the case with air pollution control and the oil embargo of 1973–74, policymakers are virtually forced to reassess the goals and means of clean air policy. They must decide whether or not events necessitate a retreat from clean air policy objectives to secure a forward-moving economy.

Our assumption is that policymakers react differently to such a challenge. In an institutionally competitive situation, with an upsurge in public demand and with high visibility in the public eye, the individual policymaker will be inclined to respond quickly and dramatically to new stimuli. This is even more so if the policymaker will not be responsible for implementing his recommendations for policy change. Under such circumstances, his strategic calculus will lead him to stress popularity more than continuity.[18] In a less competitive policy context, with less visibility in the public eye, the individual policymaker will be less inclined to recommend dramatic policy changes. He will also consider alternatives for which he could take responsibility at the stage of implementation. His strategic calculus will stress continuity more than popularity. The policymaker in the former context will reach out for the desirable, while a policymaker in the latter context will go for the feasible. Already tired by the race, the hare may be led astray by new stimuli, while the tortoise continues along the road with only slight modifications in speed and direction.

Again and again, we came back to the importance of the political context to policymakers' considerations of policy choice and change. In this respect, our study is closely related to a couple of scientific propositions that can be inferred from results obtained in recent comparative studies of environmental politics and policy:[19]

1. The more open and conflict-oriented the political system, the more immediate and substantial the response to problems of environmental quality but the less substantial and successful the implementation of adopted policy alternatives; and

2. the more closed and consensus-oriented the political system, the slower and more incremental the response to problems of environmental quality and the more deliberate and successful the implementation of adopted policy alternatives.

However, the studies from which these propositions are derived are neither as well entrenched in the causal mode of analysis as the formulation of the propositions may suggest nor consistently cast in the intentional mode of analysis. By adopting the latter mode of analysis, this study tries to find out whether the United States corresponds to the first and Sweden to the second of these propositions. It is important to remember that the study does not assume that system characteristics as such cause differences in policy choice and change, but only that they present policymakers with quite different political contexts of choice. Since we assume that policymakers are rational actors, capable of ordering policy alternatives according to their preferences, it follows that it is through an analysis of policymakers' perception of problems, calculation of consequences, and stating of preferences that we may be able to explain the differences in policy choice and change in the two countries.

Before proceeding further, we should make some comments on the latter part of each of the two propositions. They indicate that we can expect two outcomes of policy implementation, one unsuccessful and one successful. In other words, if there is a close connection between implementation and impact, we would end up with two different types of clean air.

However, the assumptions underlying the propositions may not be valid. One assumption seems to be that the contexts of choice and change are the same in policy implementation as in policymaking and another that changes in policy are followed by similar, if not necessarily simultaneous, changes in implementation.

The United States case is especially important here, since the courts have such an important position in the context of implementation. The citizen may take administrative decisions to court to have them assessed in terms of their consistency with legislative intent. The courts have a very special context of choice. Their decisions are restricted to the substance and preference calculi discussed earlier. This means that they must limit themselves to the intentions of the legislation (i.e., the policy preferences) and to the substance criteria outlined in legislation. As long as the original policy preferences prevail, the courts will continue to issue orders outlining what administrative action should be taken to implement the policy. Since the implementing agency is bound to follow the court's opinion, the original policy may continue to be implemented long after policymakers and administrators have begun to make strategically necessitated moves to change it. In essence, policy implementation continues to lead its own life, thanks to the constraints and incentives provided by the balance-of-power context. One could perhaps go so far as to say that this is the essence of that system; the oscillations and swings in the legislative arena are counteracted by the time-consuming but more steady judicial process.

Comparative Policy Analysis

For those attempting comparative policy studies within the intentional mode of analysis, an important problem of comparability occurs. One faces the challenge of giving adequate representation and meaning to the arguments, justifications, motivations, and statements of intent presented by policymakers either in favor of or against different policy alternatives and different aspects of these alternatives.

This is the problem of equivalence. The problem has two dimensions: (1) the equivalent meaning of concepts and (2) equivalent implementation and measurement of the concepts. The scholar faces the task of giving concept definitions that carry equivalent connotations and meaning in the different cultures being studied. But comparability also depends on how the concepts are measured in different national settings. The analyst must decide between comparing objective activities on the one hand with subjective cultural meanings on the other. The first is attractive because measurement is made easier and comparisons more handy; it is central in comparative analyses of a causal character. The second is necessary for comparative analyses of an intentional kind, for "definitions and measurements of concepts must be *both* as objectively measurable as possible *and* take into account the meaning assigned to the concept by the actors under investigation."[20]

To overcome the problem of equivalence, this study proceeds along three lines. Earlier, the main considerations influencing the substance, preference, and strategy calculi of political actors were defined. There should be no difficulty in comparing physical-environmental, socioeconomic, technological, and political aspects of policy considerations since there is no indication that they carry different cultural meanings. As we analyze chosen policy alternatives in the following chapters—major control approaches, distribution of policy powers, and provisions for public participation—further classifications are offered. In regard to distribution of policy powers, the different policy functions and branches and levels of government available for choice are outlined. The same is done for channels and activities open to public involvement. In this way, we hope to establish definitions that include both identical and equivalent items for both countries.[21]

Finally, we also try to order and classify the arguments used by political actors to motivate, justify, or condemn policy alternatives. Arguments for or against a certain control approach, a certain distribution of power, or provision for public participation can quite easily be classified under a limited number of general headings that have similar meanings, at least in developed industrial and democratic countries.

Comparative analysis is characterized by the limited number of basic units of analysis.[22] As in all scientific effort, there is a drive toward finding generally valid relationships. But even with the most rigorous use of scientific comparison, could one claim that the findings and conclusions concerning relationships between policymakers' calculi and policy content are valid for political units other than Sweden or the United States, or even for other policies within these countries? The answer is, of course, no. Conclusions from one case and two countries do not justify assertions of a *generally* valid relationship. At most, generalizations may be carried to the politics and policies of pollution control in industrialized nations.

The conclusions may, however, be used as a point of departure for making analytic assertions about regularities and associations on a higher level of generality. Such assertions could then be used in comparing other policies within a wider range of nations.

THE CHOICE

CHAPTER 3

Practicability or Principles?

Possible Policy Alternatives

Air pollution comes from stationary and mobile sources. Power plants, factories, and homes emit pollutants into the air. Exhausts from cars and airplanes contribute to the degradation of ambient air quality. The effects of air pollution are threefold. Effects on public health include several respiratory diseases, such as lung cancer, emphysema, chronic bronchitis, and asthma. Laboratory studies with animals have conclusively demonstrated the association of air pollution with such diseases. The ecological effects of air pollution include the detrimental effect on soils and freshwater systems of sulfur both in air and in precipitation. Economic effects include not only reduced yield of products grown on soils exposed to air pollutants but also such property damage as corrosion of metals and coatings damage.

The major approaches available to governments embarking on a course of air pollution control are regulatory or distributive (see fig. 7). A regulatory approach could have two main characteristics, which can be viewed as alternatives or complements to each other. An air resource management approach (ARM) is primarily directed toward the quality of ambient air as a whole. It is by and large coupled to the effects of pollution, and one could thus distinguish three types of ARMs: (1) one that aims at a quality of ambient air where no ecological effects of pollution occur, (2) another that aims at a level of air quality where no health effects of air pollution occur, and (3) a third that aims at a quality of ambient air that precludes most economic damages from pollution. The individual source control approach (ISC) aims at regulating emissions from individual polluting sources. Even if the ISC is primarily aiming at reducing concentrations of air pollutants at points of emission, it is evidently also tied to a judgment as to what effects a certain level of emissions would have

A distributive approach to air pollution control involves charges and fees to be levied on polluters. In theory, at least, charges and fees will

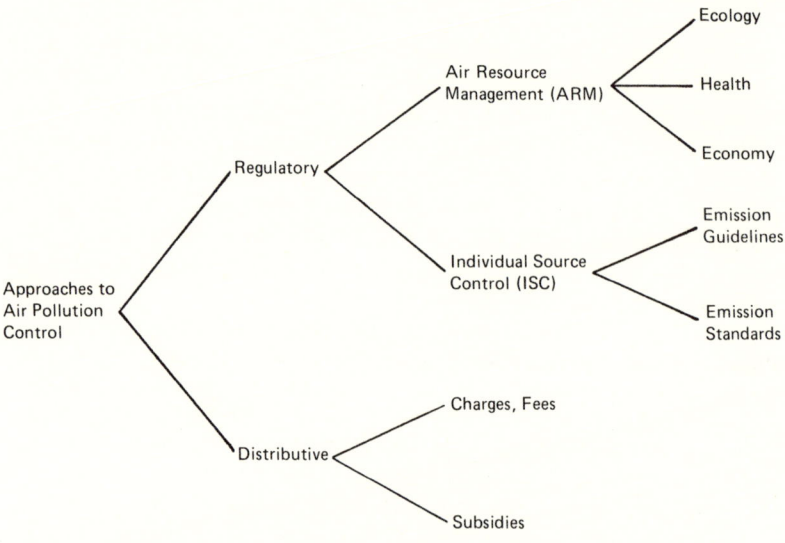

Fig. 7. A classification of alternative approaches to air pollution control

be set at levels that will give polluters an economic incentive to reduce or stop pollution. Government subsidies given to polluters for investments in pollution control equipment may also be included in the distributive approach.

Apart from these substantial alternatives, there are other, somewhat more general ingredients and issues in a situation of policy choice. How should the choice proceed? At least two main alternatives can be outlined. Policymakers may first determine the goals and objectives and then start looking for the means to be used or developed to reach the goals. They may also start with a careful look at what means and resources are available and only then decide on policy objectives. Some policy choices may be characterized as "ends in search of means," and others as "means in search of ends."

This chapter will analyze the choices of major approach to air pollution control in Sweden and the United States. Efforts will be made to clarify not only differences in selected alternatives but also differences in the arguments and motivations used and presented by policymakers. An effort will be made to indicate whether there was any difference between the two countries with regard to the ends-means dimension. Can it be established whether the countries reached out for the desirable or settled for the practicable, and what the motivations were in each case?

Sweden Argues the Practicable

The Royal Commission on Pollution and Other Nuisances reported in 1966, after three years of work. In essence, it favored a franchise or licensing system for polluting activities, modeled on the Water Pollution Control Statute of 1956. It recommended limiting pollution as far as possible, according to criteria of technical practicality and economic feasibility. Most of the report pondered on problems of adjustment to existing legislation and administrative coordination. Of the nearly 500 pages in the report, only a handful dealt with existing air pollution problems in Sweden.[1]

The Minister of Justice used the 1966 report as a point of departure in his arguments for the 1969 legislative proposal.[2] The first step was to put the problem of environmental quality and pollution in its proper context. Socioeconomic trends, such as industrialization, urbanization, and rapid development in transportation, had made pollution more visible and acute. New technologies and structural changes in industry were increasing pollution on an ever-larger scale. The increasingly social and collective nature of pollution made it clear that solutions through individually instituted civil actions in courts were no longer practicable. Government action was necessary.

But what should be the scope of such actions? Should the principles of environmental quality and public health alone be used as guides to curb polluting activities, even though these activities contributed considerable social and economic benefits to society as a whole? Clearly, this was not the minister's view. He felt that actions to control pollution had to steer carefully between environmental costs and socioeconomic benefits and consider environmental quality as only one among many legitimate societal interests. The minister summed up his arguments by stating that "governmental actions to promote environmental quality in the stiffening competition with other social and public interests and goals must proceed along several lines, such as foresighted planning, efficient administrative organization, scientific research, and modern legislation."

What then, should be the major approach of such modern legislation? Theoretically, there are many possible alternatives with regard to its scope, precision, and toughness. One can imagine a single comprehensive law, covering all things environmental, as well as specific laws for each and every pollution problem. Legislation may range from a few very general clauses, which leave implementation in the hands of administrators, to detailed regulations, making administrators less free to use discretionary powers. Laws may be very stringent in the search for better and more efficient means, as well as more flexible in what is deemed possible.

When one examines the restrictions placed on these aspects of the legislation one finds that the Minister of Justice saw the range of possibilities as much more limited. The minister acknowledged the need for regulating nuisances and types of pollution other than water pollution. But for the moment, at least, comprehensive alternatives were out of the question. Such an alternative would break up already well-functioning bodies of law, take a long time to formulate, and thus detract from the possibility of immediate action to abate pollution. From a physical-environmental point of view, pollution problems varied considerably from place to place; socioeconomic conditions were not the same from one polluter to another. Because of the technical connection between air and water pollution, joint regulation was necessary, which was very practicable from the polluter's viewpoint. However, the approach to such joint regulations had to take into account the licensing procedure long since established in the field of water pollution control.

Summing up his arguments, the minister saw two central aspects in formulating the legislation: the law must be (1) general and (2) flexible. Generality was necessary to make the same considerations applicable to a wide variety of pollution problems and cases. Flexibility was needed to enable regulators to take into account the specific circumstances relevant to each case.

The minister's choice was a licensing or franchise system, modeled in many ways on the existing Water Act and the Water Pollution Control Statute, and covering in principle all pollution-causing activities and uses of property. Certain polluting activities were made subject to a franchise system, others to a system of advance notice to regulating agencies. As a general rule, industrial and other polluters were required to take all precautionary measures and observe all limitations on polluting activities that could be reasonably demanded. The scope of this requirement was to be determined with a view to technical practicality, the cost of pollution and pollution control measures, and the socioeconomic benefits that accrued from the polluting activity.

Thus, generality was achieved by making the law cover all polluting use of property. Flexibility was gained by allowing for a careful balance both of costs and benefits and of possibilities and constraints in each case. It is indeed striking to find how important socioeconomic and technical arguments were to the minister. Among other problems were existing small plants with limited economic resources. In such cases, only as an extreme measure would shutdowns be considered. Even greater flexibility must be allowed for such polluters, for example, a long lead time for control measures. In no way should pollution control measures be allowed to handicap Swedish industry in international competition.[3]

On the other hand, flexibility was restrained by the goal of "preventing, as far as possible, further pollution." A benchmark for the "possible" should be the "best available technology." In other words, what could be reasonably demanded to control pollution was the adoption of the most efficient techniques proved practicable in other plants of the same type, either in Sweden or in other countries. There was no question of pushing polluters over the edge by demanding measures that required the use of new and revolutionary technologies. Rather, the level of reasonable demands would be adjusted as new technologies developed and were proved practical. Even then, socioeconomic aspects would be taken into consideration for each case.

Obviously, a government that wanted to take credit for a progressive environmental policy could not leave implementation completely in the hands of the concerned agencies. Allowing for flexibility and discretionary powers could easily lead to a situation in which the cabinet found itself without criteria to judge the success of the policy and the performance of the agencies. Whether or not the minister had such scenarios in mind, he did propose a system of standards and guidelines to give coherence and consistency to the implementation of the general and flexible rules of the act. Without revealing any preferences for air quality or emission standards, he proposed that the NEPB (in cooperation with regulated industries) coordinate its work toward such a system. There would be no legally binding norms and standards implicitly applicable to *all* cases, but guidelines with the character of recommendations, adjustable in individual cases and thus consistent with the individual franchise system envisaged in the act. The closest the minister came to a consideration of health aspects in his arguments was his reference to the necessity of considering results of scientific research on pollution-caused health problems in working out the system of standards and guidelines.

In the Riksdag, Liberal party members criticized the minister's proposal for its lack of a coherent and progressive environmental policy, claiming the proposed franchise system was nothing but a basis for free negotiations, a doubtful approach lacking in legislative guidance. The Liberals wanted an explicit, long-term, coherent environmental policy, "one for which the politicians take responsibility." Policy goals should be worked out by politicians, they said, not administrators. Furthermore, the balance between conflicting interests should be struck according to explicit and coherent criteria, stated in economic terms, if possible, forcing polluters to reconsider locations as well as technologies. In short, environmental policy was seen by the Liberals as politically too important to be left to administrative decisions. Conservative members pointed to the effects of pollution on public health, and argued for more involvement of medical expertise in the decision-making process. However, these minority views were voted down in the Riksdag.[4]

In this debate, which took place in May 1969, Liberal party members expressed fears that a system of standards and guidelines could not be worked out in the near future. However, their assumption proved wrong. Throughout the spring, such a system had been discussed by the NEPB and the polluting industries. In fact, a proposal for a system of guidelines covering a whole range of polluters was presented by the NEPB only a few weeks after the Environment Protection Act was passed.

The standards and guidelines were presented in June 1969 to the joint working panels on pollution abatement. These panels had been established in preceding years as a result of a proposal by the National Air Pollution Control Council in March 1966, several months *before* the Royal Commission on Pollution and Other Nuisances delivered its report in November 1966. In fact, the council had been working closely with the commission, outlining which industries should be covered by a future system of franchises and advance notice. According to the council, it would be necessary to prepare implementative rules at an early date, both to promote a smooth introduction of pollution control measures and to prevent industrial misgivings and possible obstruction. In the words of the council: "The central environmental agency needs for its decision-making detailed knowledge about what is technically practicable to abate pollution, as well as information concerning the economic ramifications of different pollution control techniques. Industry should also have an interest in getting information about the technical possibilities to control pollution. An investigation of practicable technologies for different industrial sectors—with special attention given to the specific problems of Swedish industry—could best be carried out if the council established joint working panels including representatives from the industries concerned."[5]

Up to March 1969, ten such panels were established, several of which reported before June of that year. The groups were mainly chaired by industrial representatives, with someone from the environmental agency as secretary. Of all the representatives in the groups, only one was a specialist in environmental health; all others were either technicians or business economists by profession. The environmental agency set the terms of reference, which included (1) a description of the pollution picture for each industrial sector, (2) an investigation of technical possibilities of pollution control, and (3) an analysis of the economic consequences of improved control techniques and different control requirements. Thus, the panels were not supposed to discuss or determine specific standards or guidelines.[6]

The NEPB wanted to reserve for itself the final choice concerning implementative rules. In many ways, however, the range of alternatives was already narrowed down. In its 1966 proposal, the Royal Commission on Pollution and Other Nuisances had argued in favor of a system of emission guidelines. The commission found such a system natural, given the individual plant-by-plant approach indicated by the franchise system. Furthermore, scientific research was having difficulty determining the effects of pollution. Thus, there was no practical basis for establishing a system of air quality standards. As late as 1968, however, the Air Quality Department of the NEPB argued for a combination of air quality standards and emission guidelines, a combination that was necessary if environmental health considerations were to be taken into account in pollution control decisions. As it developed, though, the department could not gather the scientific information necessary to adopt this combined approach.[7]

Given the mission and the composition of the joint working panels, as well as the cooperative spirit of the environmental agencies, the choice left to the NEPB concerned only the stringency of a number of emission guidelines that were not legally binding. Socioeconomic arguments played a central role in this process.

Specific standards were indeed discussed by the expert panels. The NEPB announced that the standards would apply to all circumstances, and not only to the normal process of production. To placate industrial arguments that this would lead to unreasonable economic consequences, the NEPB pointed out that actual permissible emission levels would be considerably more generous than their foreign counterparts. Furthermore, Swedish standards would be less stringent than foreign ones because of the large technical and, thus, economic problems of many Swedish industries.[8]

It also developed that several standards and guidelines were moderated in the time between the June and the December proposals of the NEPB. Arguing the technical and economic impossibility of adhering to the original guidelines, industrial representatives and organizations were able to bring about less stringent emission guidelines for foundries, ferroalloy plants, sulfuric acid emissions, sulfate pulp mills, and cement plants. According to the head of the NEPB's Air Quality Department, industrial views were no doubt most important in determining the guidelines. He stated that these guidelines were not radical measures but, rather, adjustments to best practicable technologies. According to some industrial representatives, discussions in the joint panels had been instrumental in "concentrating the discussions, or negotiations, on economic feasibility, i.e., *the heart of the matter*."[9]

The views of environmental health experts corroborate our thesis that socioeconomic arguments of practicality and feasibility dominated the process. After successfully bringing health considerations to bear only on the setting of maximum emissions levels of mercury from chloralkali plants, the environmental health expert stated that "in view of how the NEPB generally works things out," health considerations had played only a marginal role in the standard-setting process. Other public health experts found the NEPB approach principally unsatisfactory.[10]

Dominant Considerations

A drift from physical-environmental and health considerations toward socioeconomic and technological arguments in favor of feasibility and practicality can be found in other policy decisions. Early in 1967, the National Air Pollution Control Council recommended ambient air quality standards fo sulfur dioxide in urban areas, explicitly referring to the possible long-term effects on public health of that pollutant. Fuel oil had become the principal means for heating homes and buildings in Sweden, and oils with high sulfur content dominated the market, partly because of the Swedish taxation system. Furthermore, less reliance was being placed on hydroelectric energy sources as fossil fuels were increasingly used for the production of electric energy. In fact, sulfur dioxide was becoming the single most important air pollutant, except for carbon monoxide. The long-term effects on the environment were no less ominous than they were on public health; sulfur in air and precipitation leads to the acidification of fresh water and soil.[11]

It was soon clear to the NEPB that the recommended maximum sulfur dioxide levels in ambient air were continuously exceeded in urban areas during the winter season. Obviously, something had to be done, but what? Several means of reducing sulfur dioxide levels could be envisaged, such as desulfurization of fuel oils, stack gas cleaning, setting a ceiling on sulfur content in fuel oils, and changing the tax system to stimulate a switch to low-sulfur oils.

In March 1968, the NEPB wrote to the Minister of Agriculture, arguing for adoption of the third alternative. Pointing first to the serious effects on public health and environmental quality of increasing sulfur dioxide pollution, the NEPB emphasized that if present trends continued, emissions of sulfur dioxide would increase 2.5 times in the next seven years. Only if total emissions were kept down could the ongoing acidification of fresh water and soils be halted. The agency proposed a time schedule for deescalating sulfur content from 2.5 percent weight in 1969 to 1 percent weight in 1974. The NEPB stated that its proposal would be cheaper than both desulfurization and stack gas cleaning. Furthermore, it was immediately applicable, while desulfurization and stack gas cleaning needed technological developments. The NEPB pointed out that reducing sulfur content to 1 percent necessitated desulfurization capacity. If the NEPB plan were adopted, it would require large investments up to 1972.[12]

Industry agreed with the necessity to fight air pollution. However, it pointed out that the NEPB proposal would have a negative impact on the international competitiveness of Swedish industry by driving up its costs of production. Investments in desulfurization plants would amount to 1.3 to 1.4 billion Swedish crowns up to 1975, and the Swedish Petroleum Institute indicated that better knowledge of desulfurization technology was necessary before launching such an investment program. Full capacity would not be reached by 1975. Given the limited availability of naturally low-sulfur oil, Sweden's energy supplies could be endangered if the NEPB program were fully adopted.[13]

The minister agreed with the arguments of the NEPB concerning health and environmental quality. He even stated that pollution control costs should not be allowed to prevent necessary action. Confronted with the strong technical and economic arguments of industry, however, the minister would adopt only the first part of the NEPB program, the 2.5 percent limit for June 30, 1969. Later steps were to be taken only after the NEPB and industry had jointly investigated the technical possibilities and economic consequences of further limitation of sulfur content. Future measures, said the minister, "should be taken at the pace permitted by practical and economic conditions."[14]

In the Riksdag, Liberals and Communists argued for immediate adoption of the whole NEPB proposal. Liberals found it reasonably adjusted to technological and economic constraints, but wanted further investigations concerning the cost/benefit distributions caused by the program. Communists wanted a policy that would have "positive repercussions of a 'pushing' character." Adoption of the whole program would be a clear signal to oil producers that low-sulfur oil must be forthcoming without delay. Conservatives believed that the full NEPB program would jeopardize the international competitiveness of Swedish industry. Profit margins were already small, they stated, and this program would add additional costs to production. Steps beyond the 1969 sulfur reduction must await economic and technological investigations. And why not change existing tax laws to provide incentives for switching to low sulfur oils? To this the committee answered that it would disturb established taxation principles, put Swedish industries at a competitive disadvantage, and unacceptably increase rent and housing costs. The cabinet proposal was fully accepted by the Riksdag in October 1968.[15]

Exhaust Control Development

In November 1965, the Minister of Transportation established a specialist panel to lead a five-year program on exhaust control technology development at the State Motor Vehicle Exhaust Laboratory. Referring to the dangerous health effects of carbon monoxide and lead, the minister argued that Swedish cities with their narrow streets were "unfit for automobile traffic." But improved city planning and mass transit would not be sufficient. Automobile exhaust pollutants must also be controlled and eliminated. Because of differences in urbanization, driving patterns, and car populations, foreign control methods were not immediately applicable to Sweden. The specialist panel was to work closely with Swedish car manufacturers to provide much-needed knowledge on Swedish driving patterns and the health effects of exhaust pollutants. Its primary mission was to develop technologies for a cleaner engine and more efficient exhaust control systems.[16]

During the following two years, debates in the Riksdag focused on the necessity and feasibility of speeding up the program. There was no difference of opinion on the necessity of taking action to protect public health, but Liberals wanted to accelerate the program. They argued that since practicable technologies to bring down the carbon monoxide content were now available, Sweden should adopt the provisional California standards. Using the responsible agency's arguments, the Social Democrats argued the uniqueness of the Swedish situation, urging that clear recommendations from the specialist panel on a reliable and cheap technology be awaited to avoid unnecessary burdens on car owners. The Transport Minister said that adopting provisional standards would distort the specialist panel's program for technology development and delay statutes best suited to solve Sweden's air pollution problems. The panel was ahead of schedule, and the technological information needed for decisions on a compulsory exhaust control system would be forthcoming soon.[17]

Later in 1967, the minister presented the following arguments: The unique features of Sweden's air pollution problem prevented the adoption of provisional standards and measures modeled on California or federal United States policies. The climate, the driving patterns, and the car population made Sweden's problem one of high carbon monoxide levels during rush hours. In California, the problem was photochemical smog, which necessitated tough regulations of hydrocarbons. Furthermore, the United States approach was based almost totally on health considerations.

According to the minister, the Swedish approach should involve pollution control measures that improved environmental quality at reasonable costs and with no substantial negative effects. The minister believed such measures could be taken even if their necessity from a health point of view was not fully shown. Therefore, he proposed a legislative change to make closed crankcase ventilation systems mandatory on new cars, beginning with the 1969 models. The technology was available and practicable, and the costs to car buyers would be negligible. The reform would eliminate up to 30 percent of hydrocarbons in auto exhausts.[18]

When the specialist panel presented its 175-page report on auto emission controls in April 1968, it could point to some staggering data. Between 1951 and 1966, the number of cars on Swedish roads had increased sixfold. Concentrations of certain pollutants in ambient air were already above recommended levels in urban areas. If present trends continued, the total amount of auto emissions would double within fifteen years. Obviously, time was ripe for action.

But what kind of action, how soon, and at what costs? The panel stressed both city planning and traffic regulations, but concentrated on technical alternatives for auto emission control, such as modifying engine performance, installing exhaust cleaning and control systems, and regulating fuel composition. Focusing specifically on the second approach, the panel argued that practicable technologies were now available. All tested control technologies could meet United States requirements for 1968 models, and 50 percent of all cars sold in Sweden were of makes that, when exported to the United States, were equipped with such emission control systems. From a technical point of view, therefore, the United States performance standards could be adopted. But since cars are an internationally marketed commodity, it might be economically disadvantageous to run ahead of the forthcoming European ECE standards. Pondering over this dilemma, the panel finally found it more desirable to allow technological developments to have a swift beneficial impact on the Swedish air quality program than to await European cooperation. To lower carbon monoxide and hydrocarbon emission, the panel proposed that exhaust control systems be made mandatory beginning with 1971 models, with a gradual stiffening of requirements for 1973 and 1975 models.[19]

Most comments were favorable, but the Swedish Car Manufacturers' and Wholesale Dealers' Association found it economically insane to adopt standards out of joint with the European program. As usual, the Minister of Transportation understood that the proposals emanated from the desire to protect public health. But, also as usual, he argued that control measures must wait until a technological base for desired actions had been developed. He again argued the uniqueness of the Swedish situation. United States performance standards could not be adopted, he stated, since they had been designed for a wholly different pollution problem; neither could European standards be adopted, since they were too lenient and favored larger cars. The panel's proposal would place Sweden in the "middle of the road," by adopting a uniform standard adjusted to technical practicability for all but the largest cars. For such cars—constituting 2.5 percent of Swedish sales—the standard might even be expected to provide incentives for new technology developments. The minister argued for adoption of the first step of the proposed program, which would increase list prices by no more than 300 Swedish crowns. But the second and third step should not be adopted now; "only when we know that practical solutions are available will such measures be taken." The proposal was adopted by the Riksdag without any debate.[20]

The April 1968 report also contained recommendations for the regulation of gasoline content. Arguing strictly along the lines of technical practicability and economic feasibility, the specialist panel considered it impossible to make a swift and total shift to leadfree gasoline. To be implemented, such a shift must await a long process of developing refinery and engine technologies. Furthermore, the already limited Swedish refinery capacity implied enormous investments to make such a move. However, practicable and feasible methods of *lowering* the lead content were available, requiring only modest investments in existing refineries. The panel's proposal recognized two alternatives: (1) a gradual lowering of allowable lead content beginning in 1970 and (2) a system of tax incentives to promote the use of low-lead gasoline.[21]

While all this was going on, a Social Democratic Riksdag motion was already pending, asking for a swift and total prohibition of leaded gasoline. A group of Social Democrats wanted to put health considerations before practicability, insisting that the acute risks of lead poisoning among urban residents necessitated swift action. The technology *was* available; the costs were *not* prohibitive. Medical expertise should have more influence and "every Swedish citizen's right to clean and nonpoisonous air should be safeguarded," they said, even if health risks of leaded gasoline were not fully known. The Riksdag's Second Law Committee was positive, but said that economic and technological realities necessitated a gradualistic approach. The committee had learned that negotiations on a first reduction in lead content were in progress within the administration, and expected swift action. Such action came in January 1969, when the Poisons and Pesticides Board announced that maximum allowable lead content would be lowered to 0.7 grams per liter of gasoline, effective January 1, 1970.[22]

The United States Argues the Desirable

"Air is our most vital resource, and its pollution is our most serious environmental problem. Existing technology is less advanced, . . . but there is a great deal we can do within the limits of existing technology—and more we can do to spur technological advance." Thus began the air pollution control part of President Nixon's "Message on Environmental Quality" (February 10, 1970). The tone was one of cautious alarm. As for motor vehicle pollution, it was "quite possible that by 1980 the increase in the sheer number of cars in densely populated areas would begin outrunning the technological limits of our capacity to reduce pollution from the internal combustion engine." In highly industrialized areas, stationary source pollution could already "quite literally make breathing hazardous to health."[23]

In outlining his program, Nixon seems to have been guided by several considerations. First, he stressed industry's own capacity to solve its pollution problems. To an "encouraging extent," car manufacturers and oil companies were trying to produce a leadfree gasoline and an engine that could run on it. Nixon believed that this tendency should be promoted. Industries had also been adopting ambitious pollution control programs. But—and here lay the root of Nixon's second consideration—the present policy of establishing Air Quality Control Regions and making the states within them responsible for standard-setting would cause industry great trouble. Pollution-control spenders would find themselves at a disadvantage against less conscientious competitors. Control-minded states and communities increasingly found that desirable industries tended to locate within the boundaries of more permissive rivals. To simplify the task of industry in pollution control (i.e., to make air pollution control competitively neutral), a system of nearly uniform standards had to be established.

"Nearly" was as important to Nixon as "uniform." The latter would serve to maintain environmental quality in the face of industrial expansion and to guarantee the earliest possible elimination of extremely hazardous pollutants. The former would satisfy the needs and demands of the states within the federal system. In Nixon's words, the program could be designed "to achieve any higher levels of air quality which the states might choose to establish." Federal measures would evidently function as the baseline requirement for air quality, with states free to choose higher, but never lower, quality levels.

The president proposed several measures in accordance with these views. For mobile sources he envisaged a blend of short-term and long-term measures of sticks and carrots: (1) immediately, new performance standards for 1973 and 1975 car models, including standards for nitrogen oxides and particulates; (2) legislation to allow testing of samples off the production line throughout the model year; (3) legislation to make possible the regulation of fuel composition and additives; (4) an extensive federal research and development program in "unconventional vehicles"; and (5) an incentive to private developers (federal government purchase of privately produced unconventional vehicles at premium prices).

The president's desire for competitive equity seemed to permeate his proposals for stationary source pollution controls: (1) federally established nationwide air quality standards, with the states to prepare in one year an abatement plan to meet those standards (the plan could contain state emission standards for existing sources); (2) accelerated designation of interstate air quality control regions; (3) federal national emission standards for extremely hazardous pollutants and for *new* facilities that would be major contributors to air pollution. All standards were to protect public health and welfare, but federal and state emission standards were to be set with appropriate consideration to technological feasibility.

Nixon's proposals went far beyond what had so far been contemplated by policymakers on the Hill. Senator Muskie's own bill of March 4, 1970, seemed content with refining his own 1967 legislation. In the House, however, there was immediate bipartisan support for the president's proposals, and hearings began on March 5. Thus, the balance-of-power system provided for an increasingly competitive policymaking process. In this triangle drama, how did policy choices develop, and what arguments were used?

The House Acted First
On June 3, 1970, the House Committee on Interstate and Foreign Commerce reported favorably on a bill introduced by the members of the Subcommittee on Public Health and Welfare on April 27, only a few days after Earth Day. The tone was far more aggressive than the one found in Swedish reports. The report stated that air pollution was seen as a threat to the health and well being of the American people, and there was now an urgent need to speed up, expand, and intensify the war against air pollution to assure that "the air we breathe is wholesome once again." Responsible for the regrettably slow progress in air pollution control, according to the committee, were the cumbersome and time-consuming procedures called for under the 1967 act and the poor organization and performance of the federal administration. Thus, both Muskie and the president got their share of criticism.

But the picture was not entirely bleak. There was an enormous public awareness of the threat of air pollution to health and well-being and corresponding pressure to have "stringent controls imposed and enforced effectively at the earliest possible date. . . . This ground swell is important if we are to secure clean air everywhere in the United States, and it is important that this momentum not be lost."[24] At issue, then, was how far Congress should go in satisfying public demand for clean and healthy air and to what extent technical and economic feasibility should be allowed to influence policy.

The committee's arguments and recommendations in many ways resembled those of the president. By making the federal government responsible for setting national air quality standards, and by making each state an air quality control region, time would be saved for more direct action on pollution abatement, which would presumably serve the public interest. But the main argument for federal emission standards for new sources and extremely hazardous pollutants was still competitive justice and the fear of pollution havens. Furthermore, such standards should be set with "appropriate consideration to technological and economic feasibility."[25]

But what could the government do regarding auto emissions, which constituted 60 percent of the nation's air pollution? It was not equipped to design cars and compose fuels. Therefore, the committee hoped "that the two great industries—automobile manufacturers and automotive fuel producers—will join hands to develop the most effective technologies." But it reserved residual authority in the government to take necessary measures if the industries failed to do so on their own. The committee recommended testing auto emission control performance off the assembly line, but thought that actual use inspection should be considered only if it was technologically and economically feasible. Another indication of this interest in practicability and feasibility is that 42 percent of the authorizations were for technological research and development.

The committee even watered down Nixon's recommendations for the regulation of fuels and additives. It argued that fuel standards to protect public health and welfare must be "based on specific findings from relevant medical and scientific evidence." It must also be established that it would not be technologically or economically feasible to achieve automotive emission standards in the absence of fuel standards.[26]

Expressing additional viewpoints, three Democrats described the committee bill as woefully inadequate to meet the menace of auto pollution, which was threatening to strangle urban areas. The situation cried out for drastic remedies, and they envisaged a complete reversal of the standard-setting strategy. Instead of adjusting standards as technology developed, they wanted standards that would push technology forward. The "wails of the auto industry that meaningful improvements in their product pose insurmountable cost and engineering problems" were not impressive. Car manufacturers had been able to make special emission control systems to meet the more stringent California standards when "the public demand [was] great enough." They wanted legislation telling "the auto industry to either shape up or devise another way of powering our cars."[27]

The House debate on June 10, 1970, was relatively short. In principle, all participants supported the new legislation. Uppermost in their minds was the necessity to protect public health. They regarded air pollution as a threatening and serious health hazard and a major national danger. Members of the House seem to have been well aware of the overwhelming public concern for air quality and the ground swell for strong legislation. Furthermore, there was considerable agreement that mobile source emission was the most pressing pollution control problem. But of greater importance and relevance here is that there was a great deal of disunity concerning the basis of policy choice—"practicability and feasibility" or "public interest and health."

Majority spokesmen argued that the practicability approach was most reasonable and in accordance with the American way of doing things. There simply was no technology available at the moment to eliminate auto emissions, they said, but there was hope that there would be in three years' time. Anyway, the law gave the Secretary of HEW authority to set more stringent standards as soon as they were deemed technologically and economically feasible. These spokesmen also pointed out that the bill provided Congressional review. Congress could come back "3 years from now, [and if] progress has not been made by that time, we can take the necessary action which will be required."[28]

Dissenting committee members and several other members of Congress challenged this approach. The point was long since past, they said, "where we should allow polluters to plead that their economic interests are being threatened by clean air." Their approach was that it was not acceptable to write technological and economic feasibility into law because this would emphasize their importance over and beyond the public interest. The massive public concern for and interest in clean air, and the pressing duty to protect public health, indicated that it would be "better to set a goal and force the kind of action that will meet it." They argued that the law should require use of the most advanced technology available, regardless of whether or not a particular industry found it economically feasible. Furthermore, Detroit must be warned that it had to develop the cleanest possible energy source. Several amendments were offered that aimed at pushing technology and establishing a more clear-cut public interest and health approach. They were all defeated, but the debate indicated considerable House support for more stringent policy alternatives.[29]

The Senate Expanded Policy
Would Muskie take advantage of this situation to reestablish his leadership—and his reputation—in the environmental policy field? If so, would the Senate go along? Throughout the summer, there were rumors that the Muskie Subcommittee on Air and Water Pollution contemplated policy alternatives that went far beyond anything recommended thus far in pollution control. Its report on August 25, 1970, fully confirmed the rumors. What the Muskie subcommittee proposed was nothing less than a complete revision of the United States approach to air pollution control. During its consideration of the bill, the full Committee on Public Works came under severe lobbying and pressure, and did make some important changes. However, in the committee report on September 17, 1970, the approach recommended by the subcommittee was essentially upheld.[30]

Air quality criteria documents submitted to the committee indicated that air pollution with direct adverse health effects was more common and increasing more rapidly than generally believed. The committee's deep concern for the protection of public health had heightened as a result of this information, and caused discussions regarding the use of technical feasibility as the basis for ambient air standards. In ordering its preferences, the committee determined "that (1) the health of the people is more important than the question of whether the early achievement of ambient air quality standards protective of health is technically feasible and (2) the growth of pollution load in many areas, even with application of available technology, would still be deleterious to public health." Protection of public health—as well as environmental quality—was a national priority, and national ambient air quality goals should be set at levels necessary "to protect public health and welfare from any known or anticipated adverse effects of air pollution." There was also one more immediate objective. Air quality standards "protective of the health of persons must be achieved within the 3-year period" following the federal approval of state plans to implement national ambient air standards.[31]

The committee was well aware of the consequences of its ordering of priorities. Protection of public health had its price. It would require heavy investments in new technology and new processes, as well as changes in operating procedures and fuels. If existing sources could not meet the standards, they must be shut down. Implementing health protective standards would require changes in land use, transportation, and energy policies. A whole car generation (ten years) would come and go before the motor vehicle population would be adequately equipped to meet the proposed standards. During that time, as much as 75 percent of traffic might have to be restricted in some metropolitan areas. Highway, housing, and urban development programs would have to be changed. Administrative staffing at the federal level would have to be doubled, and supplemental funding provided immediately.[32]

The clear-cut order of priorities permeated the committee's recommendations and arguments for more concrete control measures. Time was important, and a sense of urgency was evident in the report. Strict time limits were set for the federal government to complete the designation of air quality control regions, for it to announce and promulgate ambient air quality standards, and for states to develop (and the federal government to promulgate) implementation plans for such ambient standards. In setting such standards, the federal authority should observe such margins of safety that a reasonable degree of protection would also be insured for extremely sensitive persons against hazards not yet identified.[33]

Like Nixon and the House, the Senate committee recommended federal emission standards for new sources and extremely hazardous pollutants; but its approach and arguments are interesting. With respect to new sources, the keyword was "available," not "actually or routinely used" technology. The responsible federal authority was to determine the time and cost at which a new technology was reasonably available. However, consideration of economic factors in determining availability should not affect the usefulness of new source performance standards. Neither was the authority to make technical judgments concerning how the standard should be implemented; it should only determine the achievable limits and let the polluter decide how to reach them. To top off this rather confusing line of argument the committee recommended that certification of new facilities should be granted not prior to, but at the time of, actual operation. On the other hand, it saw merit in preconstruction review to render advice to the polluter. Obviously, the committee had not clarified to itself if (and how) feasibility considerations could be kept apart from the determination of achievable limits. With respect to extremely hazardous pollutants, they said that emissions should be prohibited unless there was ample evidence that a greater than zero emission presented no hazard to health.[34]

Nowhere was the reversal in policy strategy more radical and more visible than in auto emission control. Where existing standards were built on economic and technological feasibility, the new proposal pressed for the development and application of improved technology. It also advocated standards built on the degree of control necessary to protect public health rather than limited by the degree of technology available. The committee proposed legislation that provided for auto emission standards to reduce by 90 percent carbon monoxide, hydrocarbon, and photochemical oxidant emissions, as well as emission from other pollutants. Additionally, the committee wanted legislation to the effect that these emission reductions should be reached by

January 1, 1975, for the first three pollutants and by 1976 for other pollutants. Originally, the Muskie subcommittee made the deadlines final, but the full committee put in an amendment to the effect that a one-year extension would be possible.

How did the committee argue for its radical proposal? It based its recommendations on the so-called air quality criteria documents developed by the federal administration. These documents—containing, for example, information about the character, spread, and effects of air pollutants—indicated that maximum allowable levels of pollution agents associated with auto emissions were substantially exceeded in many cities. The derivation of safety- and health-related auto emission standards from ambient air quality standards (which were, in turn, derived from air quality criteria) indicated that only the standards preliminarily proposed for the 1980 models would provide the safety margins desired. In effect, this meant a more than 90 percent reduction of auto emissions. Given the normal car lifetime of ten years, and provided that all post-1980 cars achieved the calculated emission reductions, it would thus be 1990 before the health-related auto emission standards could be reached.

Still the committee recommended the tough 1975-76 deadlines. The rationale for this choice is very brief in its report. The committee concluded that since the responsible federal agency's "advanced power system research and development program, as currently planned, is structured to provide, by 1975, two second-generation prototypes capable of meeting the 1980 emission-reduction goals," and since the normal lead-time for technological innovations (according to the manufacturers) was two years, the industry would have time enough to produce engines that would meet the deadlines."[35]

The committee also proposed continued prototype testing and added proposals for compliance testing of all cars in use. Each vehicle was expected to conform to emissions standards for 50,000 miles and each manufacturer to warrant the performance of emission control systems for that lifetime. The committee based its proposal on consumer protection arguments: the bill would require Americans to make substantial investments in emission control systems. Such investments would be defensible only if the systems conformed to emission standards for the lifetime of a vehicle. Otherwise, potentially "billions of dollars of consumer investment would be to no purpose." Finally, the committee proposed that the responsible federal authority should have the right to register fuels and to prohibit or control fuels that produced emissions directly endangering public health.[36]

What was remarkable about the proposal was the strict timetable laid down for implementation of the act, as well as the different degrees of administrative discretion provided. National ambient air standards as well as state implementation plans were to be published and promulgated within specified time limits. Furthermore, state plans were to be such that the ambient air standards could be achieved within three years after promulgation. Deadlines for auto emission standards could be suspended for only one year, and that suspension could be granted only under certain very carefully specified circumstances.

The bill recommended that the federal agency should "to the maximum extent possible within the time provided, consult with appropriate advisory committees, independent experts and Federal departments and agencies." With respect to emission standards for new sources, the bill went so far as to prescribe preconstruction review to give the individual polluter assistance and advice. On the other hand, the committee seemed to presume that nothing but the desire to prevent new air pollution problems should guide the federal agency's standard-setting. Economic and technological factors relevant to the implementation of standards were not to be part of the final judgment concerning the availability of a certain technology and thus the setting of a particular standard. (Of course, the mutual relationship inherent in preconstruction reviews could presumably let such considerations in through the back door.) The notoriously equivocal discussion on this point indicated that federal authorities would have considerable discretion in this type of implementative decisions, and that negotiations over the feasibility of standards might occur in the implementation process.

Administrative discretion was very limited with regard to the setting of ambient air quality and mobile source standards. Since only health considerations would be allowed to influence decisions, a key role would be played by the air quality criteria. Such criteria were to be, or were already, published for specific air pollutants, indicating the concentration levels at which adverse health effects of the pollutant occurred. Once criteria were published, it would thus be easy to derive ambient air quality standards that provided desirable margins of safety. And, as the committee's discussion indicated, auto emission standards could as easily be derived from these national ambient standards. Seemingly, this interlocking system of standards left little room for considerations of technological or economic feasibility. As we shall see, however, the federal power to extend the auto emission deadlines on exactly these grounds was to be exercised several times in the mid-1970s.

Muskie's preferences and arguments were clearly spelled out in his opening statements during the Senate debate. He pointed out that the bill was being considered in a year of environmental concern in which young and old together marked Earth Day in April and in which Congress was considering an unprecedented number of environmental bills.

Muskie perceived the bill as a test of the Senate's ability to provide leadership in a time of crisis and in the face of public concern and corporate resistance. Even if the underlying philosophy of the 1967 act were valid, he said, the costs of air pollution could be counted in death, disease, and debility. Air quality criteria documents indicated that the problem was more severe, more pervasive, and growing faster than had been thought because, under existing law, considerations of technological and economic feasibility had been used as arguments to compromise the public health.

As Muskie put the problem, it was clear that nothing less than a reordering of national priorities would solve it. In his view the present bill proposed quick and drastic actions and reflected the fact that Congress's first responsibility was to protect public health rather than to make technological and economic judgments. In Muskie's own words:[37]

> Predictions of technological impossibility or feasibility are not sufficient to avoid tough standards and deadlines, and thus compromise the public health. The urgency of the problems requires that the industry consider, not only the improvement of existing technology, but also alternatives to the internal combustion engine and new forms of transportation. . . . Detroit has told the nation that Americans cannot live without the automobile. The legislation would tell Detroit that if this is the case, they must make an automobile with which the Americans can live.

The approach implied two problems. First, would industry be able to comply with the requirements and the three- and five-year timetables of the bill? Second, who would bear the costs of this radical policy change? Muskie was convinced that industry could comply. It had been presented with challenges in the past (e.g., putting a man on the moon in a decade) and met them successfully. Furthermore, the committee had done everything to allow the lead time considered necessary by the industry: two years for research and development and two years for application in mass production of the technology required.

Most other senators expressed their support for the objectives of the legislation. However, opinion was divided on the question of whether it would be possible for industry to come up with the required technology in five years. By far, the principal and most thorough criticism came from Senator Griffin of Michigan, the heartland of the auto industry. He regretted that the decision had to be taken on a too political basis. As he put it, when "everyone is for clean air and against pollution," it was difficult to propose anything that would be characterized by the press as weakening the bill. Nevertheless, he wanted to debate what he called the questionable premises of the bill. First of all, auto pollution was not the largest or most dangerous source of air pollution. Measured by amount, it contributed only about 40 percent.

Furthermore, if one measured different air pollutants by harmfulness, auto pollution's share would drop to only 12 percent!

Griffin got Muskie to admit that no discussion had taken place with the auto industry on the deadlines of the bill. This meant the deadlines were totally unrealistic. The industry would not even have the three and one-half years normally needed to produce a new car model. Procedural rules of the bill would compress the time available for inventing the totally new technology required for emission control to eighteen to thirty months. And that for a technology that had withstood breakthrough efforts for more than fifteen years! As Griffin put it, "In short, Mr. President, this bill holds a gun at the head of the American automobile industry in a very dangerous game of economic roulette . . . we could be headed for an economic and transportation crisis in 5 or 6 years time." Excessively penalizing this vital industry could jeopardize 15 million jobs throughout the country, and would affect one sixth of the gross national product. Griffin recommended that politicians stop playing scientists and technicians, and listen to the real experts in trying to assess what was feasible and reasonable in this field.[38]

Senator Muskie managed to get the bill through with only minor changes. The conference to work out a compromise between the House and Senate versions was subject to intensive lobbying and media coverage. However, the conference report approved by both houses on December 18, 1970, differed only marginally from the Muskie subcommittee proposals of September. Two weeks later, when President Nixon signed the Clean Air Act Amendments of 1970 (with Muskie, the main architect, not present), he confirmed a policy redirection and change no one could have imagined when the process started in December 1969.[39]

Summary

In Sweden, the process of policymaking for air pollution control lasted for six years, from 1963 to 1969. No policy existed previously, and the process was highly fragmented and incrementalistic. One issue at a time was addressed, and no complete look at the whole problem of Swedish air quality was attempted during the process of choice. Throughout the process, the center of gravity of policymaking lay within the executive branch of government. Cabinet ministers, Royal Commissions, and the joint working panels of administrative and industrial experts were the main participants in the process. Public concern and public reactions played only a minor role.

With respect to policy content, the Swedish policy objective never changed. Throughout, the aim was seen as controlling air pollution as far as was technologically practicable and economically feasible. The individual source control approach was derived from the feasibility criterion, to be implemented through a system of source licensing and administrative emission guidelines. The limited steps taken to control sulfur dioxide and auto emissions were also motivated by the criteria of feasibility and practicability. Since there were no ambient air quality standards or goals, no timetable was established.

In the United States, the process of changing the existing air pollution control policy was completed within one year. Between December 1969 and December 1970 a complete, comprehensive overhaul of the whole air pollution control problem was effected. The policymaking process was highly competitive, with initiative shifting between the president and the Congress and, within the Congress, between the House and the Senate. As things developed, the center of gravity settled in the legislature, in the Muskie Senate subcommittee. Public concern was of utmost importance to the policymakers, while the regulated interests (e.g., the auto industry) had less than usual access to the process.

The United States air pollution control policy experienced a major shift in policy emphasis, from technical and economic feasibility to protection of public health. From the health criterion were derived not only national ambient air quality goals and standards but also auto emission standards and national emission standards for new sources and for extremely hazardous pollutants. Very strict timetables for achieving the air quality and auto emission standards were explicitly included in the legislation.

With some exaggeration, one could characterize the Swedish approach as one of choosing an adequate objective for the available means, while the United States approach was one of going beyond available means to establish a new, and seemingly absolute, policy objective.

CHAPTER 4

Allocation of Policy Authority and Responsibility

Concepts and Arguments

A policy consists of many different but interrelated and interdependent activities, each of which must be performed by some branch or level of government to secure a successful and effective policy development. The allocation of authority and responsibility for different policy activities is therefore a central part of all major policy choices.

Policy is here seen as encompassing three main categories of action: policymaking, policy implementation, and policy review. Implementation is viewed as comprising three interrelated but distinct activities: (1) policy interpretation, involving the translation of legislative intent into standards and guidelines; (2) policy application, meaning the use of standards and guidelines to decide individual cases, such as the granting of permits, franchises, and certifications; and (3) supervision and enforcement, involving the more or less continuous monitoring of compliance with standards, guidelines, instructions, and conditions attached to individual decisions.

Policy review may also be divided into distinct activities: (1) the oversight and reassessment of policy principles and objectives (of general legislation), and (2) the review of whether or not policy implementation is consistent with general policy goals and principles, (with legislative intent).

Policymakers are faced with the task of assigning these responsibilities among different branches and levels of government. One major concern may be to find an acceptable balance between the branches of government: the executive, the legislative, and the judicial. Another is allocating authority among different levels of the same branch: the national, the regional, and the local. Altogether, this gives policymakers a large number of alternatives and possible models to follow when allocating authority and responsibility for a certain policy. The choice is heavily dependent on what values and arguments the policymakers view as central (see fig. 8). Policymakers with a preference for effective and consistent policy implementation will most likely come out in favor of a

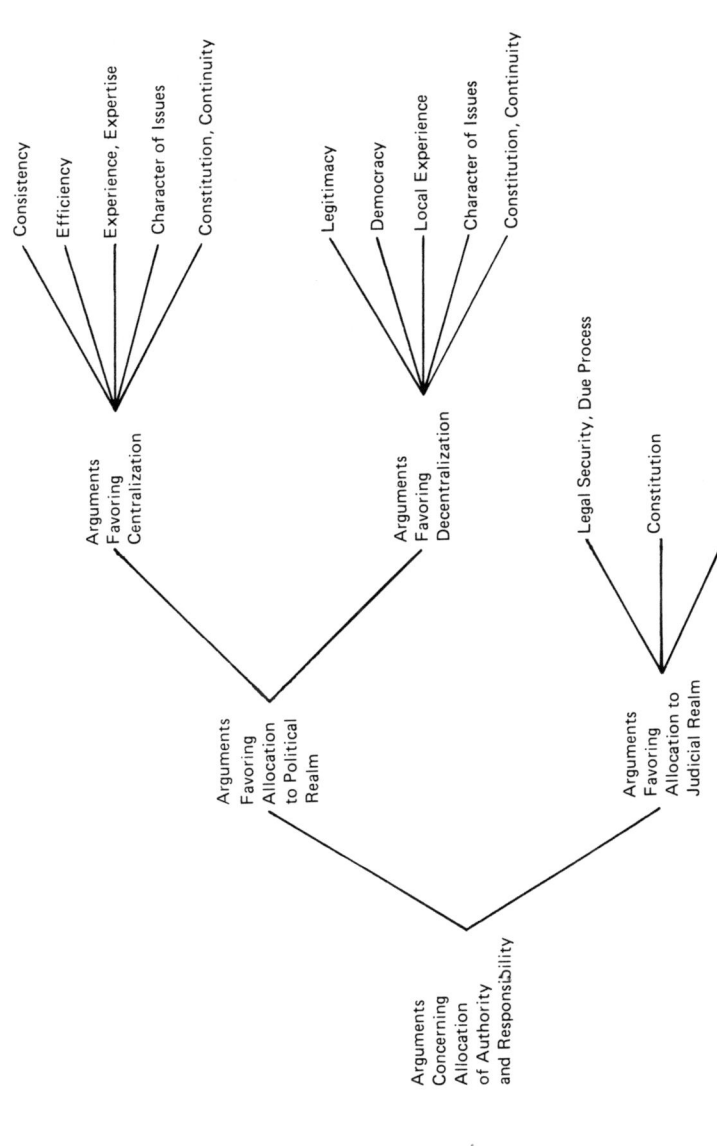

Fig. 8. Classification of arguments concerning allocation of policy authority and responsibility

pattern of power allocation quite different from that proposed by a policymaker most interested in constitutionality and due process.

Centralization, in the sense of concentrating policy authority at the highest government level, is frequently found appealing as a means to secure effectiveness, performance, and goal achievement. Centralized authority is seen as conducive also to policy consistency: a consistency not only between legislative intent and implementative standards but also among individual decisions based on these standards. Advocates of centralization say it brings together specialized expertise and develops a common core of experience and competence, all of which are conducive to performance and effective goal achievement. Centralization may also be supported for other reasons. Constitutional rules may favor such alternatives, and existing patterns of centralized policy authority may be defended as necessary to secure continuity in policy development.

Decentralization, in the sense of assigning policy power and responsibility to different branches and levels of government, is often supported as a means to strengthen democracy and enhance the legitimacy of a particular policy. Decentralization takes decisions to the people, which may be viewed both as a democratic value in itself and as a means to increase participation. Closeness and participation are intimately related to the legitimacy of a policy; consequently, decentralization is supported with arguments of legitimacy. Support of decentralization is also based on the value of involving local knowledge: policy decisions should take stock of local expertise and local knowledge of conditions, especially in policy implementation. In the case of air pollution control, variations in local socioeconomic conditions and the local pollution picture may serve as motivations for decentralization of policy responsibilities. Finally, decentralization and dispersion of authority may be called for by constitutional arrangements, granting states and local communities certain responsibilities for public policy.

What could be the main motivations for allocating responsibilities to the judicial branch? Conformity to norms of due process and the rule of law seem to be central. It is not suitable for a system of government by law to have standards of conduct and their reasonableness in individual cases decided with finality by one and the same agency. Citizens have their rights to due process and the rule of law, which require courts or at least quasi-judicial agencies. By this line of reasoning, standards should be set by legislatures or administrative agencies, while granting concessions, permits, and licenses are activities more suitable for courts. Thus, the character of the issues and decisions may call for judicial settlement.

Of interest here is finding out what role considerations of this kind played in the choice among alternative arrangements for allocation of clean air policy powers. In one sense, the commonness of air as a resource might well be used to support centralization at the expense of constitutional and existing arrangements. Centralization could be advocated as a means of avoiding pockets of no, or less stringent, regulation. On the other hand, variations in the air pollution picture, whether caused by natural or man-made conditions, might serve as an argument for spreading policy responsibility to regional and local levels of government.

Sweden Argues for Central Administrative Authority

To put the 1966 power allocation proposals of the Royal Commission on Pollution and Other Nuisances in a proper perspective, several circumstances should be pointed out. (1) The commission had in fact proposed a system of licensing and franchising of all polluting activities emanating from the use of real estate; before, such a system had only been in force for certain cases of water pollution. (2) A system of courts, boards, and agencies with jurisdiction over different parts of the area to be covered by the new environmental legislation already existed or was in the process of being created. (3) Any system of licensing and franchise involves issues of efficient policy implementation and the status and treatment of the clients involved. In short, the problem for the commission seemed to be the following: How does one establish a system of power allocation that is consistent with the character of the proposed legislation as well as with existing or upcoming institutional arrangements and that satisfies demands for both efficiency and due process?

Uppermost in the commission's mind was the need for a comprehensive examination of all polluting effects of property use. The close connection between water and air pollution pointed toward a system where the permissibility of all types of nuisances arising from property use should be examined at the same time. The licensing system proposed by the commission was of this comprehensive character.

But who should be responsible for granting licenses? Here, the commission argued on grounds of a precedent. The Water Courts already were responsible for granting permits to activities involving construction in, or pollution of, Swedish waters. They thus had the expertise and experience necessary for such franchise decisions. Furthermore, the Water Courts in general had precisely the kind of judicial, technical, practical, and economic experience needed to try the permissibility of all other types of pollution covered by the new act. Finally, the six Water Courts were regional bodies. Their geographic areas of jurisdiction were large enough to provide varied experience and expertise but, at the same time, small enough to provide judges with a valuable knowledge of local conditions. The Water Courts had indeed established a very good rapport with local industry and local opinion.

The proposal to give the Water Courts exclusive jurisdiction over licensing and franchise as well as other judicial issues, such as compensation, rested on two fundamental assumptions: (1) judicial and administrative aspects of pollution control decisions could be analytically distinguished and practically and institutionally kept apart and (2) there would be a central authority responsible for policy interpretation, supervision, and control.

Such a central authority was just about to be established at the time of the commission's proposal. In 1965, the Riksdag had decided that a central environmental agency should be established by July 1967. This new National Environment Protection Board, the NEPB, would assemble in a single authority the scattered units responsible for air and water pollution control, conservation, and noise abatement. The NEPB would thus be responsible for policy interpretation, supervision, and control for most of the area covered by the proposed environmental legislation.

Why not give it the responsibility also for granting licenses and permits for polluting activities? In answering this question, it was clear that the commission had made up its mind concerning the first assumption. Judicial and administrative issues could indeed be distinguished, and should be kept apart institutionally. Franchising and compensation were closely connected judicial issues, involving fundamental questions of fair trial and due process. This was the principal reason for letting the responsibility for these issues rest with the Water Courts. Considerations of conformity to precedents were important. The commission did not find it relevant or necessary to discuss the alternative of administrative licensing, "since it would distort the existing system for water pollution franchise."[1]

This choice still left the commission with the problem of effectiveness and efficiency. Court proceedings are time-consuming and may delay implementation. A central concern of industries is to get decisions as quickly as possible. Both effectiveness and efficiency favored an administrative licensing system. However, the commission tried a different solution. It proposed that a formal license should not be mandatory in all pollution cases, but only when pollution would have large ramifications or where the case would be of principal interest. For many cases, licensing would be optional. The polluter could go to the administrative agencies and ask for an exemption from the licensing requirement, and the agency would then decide the conditions under which the polluting activity could be continued. In this way, the great majority of cases would be dealt with expeditiously, without detracting from the effective implementation of policy objectives.

A condition attached to the adoption of this dual franchise system was that the NEPB would be given responsibility for granting exemptions. The central agency, so the commission reasoned, would have the expertise necessary for effective air pollution control. Only that agency would have the experience necessary to negotiate with and understand the problems of all types of air and water polluting industries. Later, as the resources and experience at the regional level increased, a delegation of exemption powers should be allowed to take place.[2]

As for the division of power and responsibility within the administrative branch, the commission proposed that the NEPB should be responsible for policy interpretation, that is, for setting standards and issuing guidelines. The NEPB and the state regional boards should share responsibility for supervision and control, with the NEPB as the leading and coordinating body and the regional boards carrying out day-to-day control and supervision. As before, enforcement action should be a responsibility of the state regional boards, which should cooperate with public health committees in the local communities. However, these public health committees found themselves stripped of most of their power in the proposal. Earlier, the committees had been responsible for granting permits to local polluting activities. Under the new legislation, this authority would be vested in the Water Courts and the central NEPB.[3]

Administrative vs. Judicial Licensing
The commission's proposal to split policy implementation into judicial licensing and administrative supervision and control was criticized in several comments from agencies and interest groups. They wanted a system of administrative licensing. Since this alternative had not been investigated thoroughly, Minister of Justice Kling ordered the NEPB to make such an investigation and to report on the organizational consequences of this alternative solution.

In its report two months later, the NEPB proposed an administrative licensing procedure. Franchise decisions, the NEPB argued, are by nature administrative, not judicial. A certain degree of nuisance has to be tolerated in every modern society. But that degree is the result of a trade-off between costs or nuisances on the one hand and benefits on the other. Such trade-offs are administrative, not judicial, in character, founded as they are on considerations of long-term technical, economic, and social impacts. In other sectors of social activity, such decisions are usually made within the administrative realm. In most countries, added the NEPB, such large scale environmental/economic trade-off decisions are made by administrative agencies.

The main function of the court system is to decide what is right in a particular case, while administrative agencies should actively implement the goals and policies decided upon by the government. The NEPB continued: "Now that environmental policy is to be expanded, it is evident that the center of gravity should lie in some administrative agency and not in the courts."

But in which agency? The NEPB had no difficulty finding the answer. For all practical purposes, the NEPB itself was the best choice. It was responsible for overall, long-term environmental planning, and would thus have the necessary knowledge and overview of all aspects relevant to license decisions. From the viewpoint of consistency, it would be essential that such decisions would be taken by one and the same agency instead of by six different Water Courts. Since the NEPB already had assembled essentially all the national expertise on environmental quality and air and water pollution control, the agency had the capacity for swift and effective decision making on permits and licenses. The comprehensive review so much desired by the Royal Commission could indeed be conducted more efficiently and thoroughly by the NEPB than by the Water Courts.

However, the problem of efficiency remained. If administrative licensing were to take into account all relevant socioeconomic circumstances, it might be as time-consuming as court proceedings. The NEPB therefore joined the Royal Commission's proposal for a system of mandatory and optional franchising. For certain types of plants, the NEPB should have the power to grant exemptions from the duty to apply for a license and to determine the precautionary measures and other conditions under which the polluting activity could be carried out. For other installations, the license procedure would be mandatory, and the most serious cases should be licensed only after an investigation and decision by the cabinet.

The NEPB suggested that the majority of franchise decisions be made within the organizational framework of the NEPB itself. This increased workload required increased resources and internal reorganization. Two partly conflicting needs must be satisfied, effectiveness and due process. The NEPB discussed two alternatives: (1) a separate, administrative court-type franchise board liaisoned with the NEPB and (2) a franchise board fully integrated with the NEPB. The former would satisfy the criterion of due process and rule of law, but it did not guarantee effectiveness. The franchising unit would have to be in close contact with the activities of the NEPB, and the NEPB must have some influence over licensing decisions because of its responsibility for overall environmental planning. The NEPB thus suggested the establishment of a franchise board fully integrated with the NEPB. Members would be the NEPB Director-General, an experienced judge, and the NEPB official whose area of competence was relevant to the license decision. Ad hoc groups within the NEPB would investigate each franchise case, and give advice and counsel to the franchise board.[4]

The ramifications of this proposal were indeed staggering. It meant that the NEPB would (1) give counsel to the franchise board with regard to precautionary measures and other conditions that should be attached to the franchise; (2) be a party in franchise board proceedings because of its responsibility for environmental policy implementation; and (3) function as the judge in franchise decisions, because of the NEPB majority on the franchise board. Clearly, the proposal invited criticism from the viewpoint of due process and rule of law. The comments focused on this aspect; franchising and licensing should be regarded as administrative decisions, but there must be guarantees that due process and impartiality would prevail. On this ground, the majority of agency and industrial comments disapproved of the NEPB, and recommended some form of independent franchise board.[5]

Minister of Justice Herman Kling agreed with the commission's view that licensing must involve a comprehensive examination of all kinds of pollution in each case. He also agreed with the NEPB that a permit decision is administrative, as it involves judgments according to the criteria of technical, economic, and social "reasonableness" and has important effects on policy implementation. However, the minister believed the problem was practical rather than theoretical: "It is important to find a solution that integrates the franchise system with an effective and socially beneficial environmental policy, allows for swift and efficient decision making with experienced attention given to all affected public interests, and allows for a coherent and consistent policy implementation. At the same time, all reasonable demands for due process must be satisfied."

Evidently, an administrative rather than a judicial solution would be most appropriate. The minister repeated all of the main arguments in favor of a centralized, administrative licensing system. It would be efficient; all necessary expertise would be under the same roof. The minister of justice was firmly convinced that through such a system, effectiveness and consistency in policy development would be enhanced; the central agency would have the experience, expertise, and overview necessary to make the franchise system become part and parcel of a coherent and goal-oriented central environmental policy.

But the NEPB proposal had one main weakness. The cornerstone of an effective and efficient environmental policy would be good faith and cooperation between the central environmental agency and the polluters. But the arrangement proposed by the NEPB could foster feelings among franchise applicants that due process and impartiality might not prevail. If the legitimacy of environmental policy decisions were in doubt, agency-client cooperation would be jeopardized, and with it the possibilities of a successful environmental policy development.

The Franchise Board for Environment Protection

The minister then proposed an elegant compromise. To satisfy demands for efficiency, the system of optional franchise should be enlarged to include almost all polluting activities. The NEPB would have the authority to grant exemptions from the franchise duty, and to prescribe precautionary measures and other conditions for the continuance of the polluting activity.This would increase efficiency and facilitate swift decision making; at the same time consistency would be enhanced by the concentration of authority in the central agency. The minister foresaw that exemptions would be the central element in the franchise system. To satisfy demands for due process, the minister proposed the establishment of a totally independent Franchise Board for Environment Protection (FBEP). The FBEP would function like an administrative court. If polluters were not satisfied with the NEPB's exemption decisions, they could apply to the FBEP for a permit. In turn, the FBEP franchise decision could be appealed to the King in Council. The impartiality of the FBEP should also be visible in its composition. The chairman must be an experienced judge, and the other members should be highly qualified experts with technical, environmental, and economic experience.[6]

The only authority left for the courts would be to decide on issues of compensation. Supervision and control were to be organized in the way proposed by the Royal Commission, i.e., with central coordination and regional action. Enforcement would mainly be a regional responsibility. Franchise board decisions were to take precedence over decisions made by local public health committees. Since exemption decisions made by the NEPB would not be legally binding, they could be overruled by decisions of local public health committees. To prevent jurisdictional trouble, the minister proposed a consultation procedure, involving state regional boards and local public health committees, in all NEPB exemption decisions.[7] From then on, the debate focused on the proper relationship between the FBEP and the NEPB. The minister, Mr. Kling, and his fellow Social Democrats in the Riksdag's Third Legislative Committee persisted in arguing for a close personal connection between the two agencies. On the other hand, the bourgeois opposition parties joined the cabinet's Legal Advisory Council in arguing for a completely independent FBEP.

The council stressed the demand for impartiality and due process. It is quite obvious, said the council, that in most licensing cases the NEPB has the status of a party vis-à-vis the franchise applicant. To allow the NEPB director-general or some other leading NEPB official to be a member of the franchise board would indeed create doubts concerning the FBEP's impartiality. The council concluded that any NEPB membership in the FBEP should be explicitly ruled out.

Opposition parties all proposed an independent FBEP. In formulating the minority view in the committee report, it was clear that they emphasized impartiality and due process over and above efficient and consistent policy implementation. Several ways could be found to satisfy the demand for environmental expertise in the licensing process without having NEPB officials as FBEP members. There was no question of denying the NEPB access to the licensing process; it would continue to make investigations, give advice, and issue recommendations to the franchise board. It should not, however, actually make FBEP decisions; this would be inconsistent with the norms of due process.[8]

The Social Democrats persisted in emphasizing consistency and effectiveness over and above the norms of due process. Minister Kling summarized the majority view: "The most important thing is to have the FBEP and NEPB decisions and activities coordinated to a consistent environmental policy. . . . The main reason for allocating responsibility for the licensing system to central agencies is the necessity to secure a *comprehensive, consistent, and effective implementation of policy*." The demand for impartiality and due process could be satisfied by appointing independent environmental expertise to the FBEP. The cabinet proposal was accepted after roll-call votes in the Riksdag Chambers.[9]

Of somewhat greater interest here is the debate evoked by the Liberal proposal to develop a long-term program, or plan, for the development of Swedish environmental policy. The Liberals proposed that since many issues and problems would arise in the future, the policy program should include such things as (1) the development of criteria and quality goals for different parts of the physical environment; (2) investigations of the costs and effects of different regulatory and distributive means, as well as pilot experiments; (3) the development of an environmental research program; and (4) a closer, more detailed organizational coordination of such activities as physical planning, housing and urban development, and environmental protection. The Liberals found it extremely important that this policy planning should be the responsibility of elected politicians; a Royal Commission with parliamentary representation should be appointed. Liberals argued that this program involved issues and trade-offs of an inherently political character; such decisions should not be taken by administrative agencies.[10]

The NEPB commented that it could not find any need for the kind of political policy planning envisaged by the Liberals. Sweden already had such a program for environmental protection, it stated, consisting of all environmental acts, ordinances, and agency instructions passed during the latter part of the 1960s. That program would be more efficient than the one proposed by the Liberals, which would have to be very general, containing principles already agreed upon, and thus not an adequate vehicle for future policy development. The Social Democrats agreed that planning and future policy development could be entrusted to the NEPB, whose planning and coordinating activities should form the basis for environmental policymaking.

The Liberal-Center minority view questioned this distribution of policy responsibilities. Policymaking is indeed an inherently political activity, they argued, and existing laws, ordinances, and agencies mark only the beginning of Swedish environmental policy; future developments must come through a lively political debate and review of alternative solutions, not through administrative regulations and decrees. Mrs. Anér, an outspoken Liberal environmentalist, launched principal criticism against the majority view in the committee report. The new Environment Protection Act, she said, lacked precisely what the majority said it contained, i.e., a coherent, long-term program for environmental policy development. She continued: "Yes, we would like to have—that's it—a long-term and goal-oriented environmental policy. But that policy must be one for which the politicians bear the responsibility. . . . You cannot leave the responsibility for the main principles in such an important and crucial field as environmental policy in the hands of an administrative agency, however competent that agency may be." But she argued in vain; the Liberal-Center demand for a policy-planning Royal Commission was defeated in the Riksdag by a wide margin.[11]

The United States Argues for Increased Federal Assistance

The power allocation envisaged in the 1970 Clean Air Act Amendments cannot be understood without some knowledge about the content of earlier United States air pollution control legislation. The Air Quality Act of 1967, itself an amendment to the Clean Air Act of 1963, still left the states with the primary responsibility for air pollution control. The secretary of HEW was given authority to define atmospheric areas of the country and, after consultations with state and local officials, to designate within eighteen months air quality control regions embracing whole states, intrastate regions, or interstate areas. The secretary was also responsible for formulating and issuing criteria of air quality based on the health and welfare effects of each major air pollutant. The states were then responsible for setting ambient air quality standards based on federal criteria for any air quality control region in the state, and to submit an implementation plan for enforcing these standards. If the states failed to establish the standards within a specified time period, or if the standards were deemed unsatisfactory, federal action would result.

With respect to enforcement, the federal government could proceed against a violator of state implementation plans by giving him a 100-day abatement notice. However, if no abatement action was taken, the federal government could proceed to court action only in cases of interstate pollution. In all other cases, federal enforcement action was dependent on an official request from a state governor.[12]

In his environmental message to Congress on February 10, 1970, President Nixon stated that existing legislation had several shortcomings. Federal designation of air quality control regions, so necessary to come to grips with interstate pollution, was a time-consuming process. The tendency for adjoining states to propose inconsistent air quality standards caused further delays. Insufficient federal enforcement powers hampered the development of effective abatement programs. To begin a more ambitious national effort, Nixon proposed expansions in federal authority to (1) test the performance of motor vehicles, (2) regulate fuel composition and additives, (3) establish *national* ambient air quality standards, (4) establish *national* emission standards for hazardous pollutants and for new emission sources, and (5) seek court action to abate both interstate and intrastate pollution. This would allow the states to concentrate on developing and enforcing abatement plans.[13]

The president's proposals indicated a break with existing policy. The 1963 and 1967 acts stated that "the prevention and control of air pollution at its source is the primary responsibility of states and local governments," and national emission standards for new stationary sources meant expanding federal authority beyond the usual financial leadership and assistance. The proposal could be expected to run into some trouble on Capitol Hill. In 1967, Senator Muskie had successfully led the opposition against the Johnson Administration's efforts to introduce national standards. The question in 1970 was whether policymakers on the Hill would be led by their regard for public health—a traditional federal concern—to accept federal expansion, or would use states' rights and local variations in pollution as arguments for continued state and local responsibility.

Expansion of Federal Control
In its report on June 3, 1970, the House Committee on Interstate and Foreign Commerce found it "abundantly clear" that existing policy was inadequate, slow, and ineffective. The committee argued that "the basic strategies in the nation's war against air pollution must be developed in a unified and consistent way by the federal government," while "the implementation and enforcement of these strategies will have to be effected in every community in the United States." But this required not

only new legislation but also better administrative performance. The failure of existing policy was to a large extent a result of organizational problems at the federal level, where air pollution control was not given sufficiently high priority, and of the National Air Pollution Control Administration's lack of aggressiveness in implementing the law.[14]

The committee's recommendations expanded federal powers beyond what had been proposed by the president. The secretary of HEW was given authority to establish not only national ambient air quality standards and national emission standards for new stationary sources and hazardous pollutants, but also for aircraft emissions. Federal powers to regulate auto emissions were retained and expanded to include testing of vehicles and engines off the assembly line. Federal responsibility for standard enforcement was expanded by allowing for federal inspection of installations allegedly violating new source and hazardous pollutant standards and by shortcutting procedures for federal enforcement action in cases falling within state implementation plans. In establishing these implementation plans, states were given authority to set emission standards more stringent than those established by the federal government. This authority, however, did not extend to auto emissions; only California was allowed to establish its own state auto emission standards. Like the Nixon proposal, the committee bill carefully outlined the conditions for federal takeover of state responsibilities for developing the implementation plan.

The committee presented very few arguments in favor of these recommendations. However, the few there were made it clear that consistency and efficiency rather than constitutionality were the committee's major concern. Nationwide air quality standards meant that "the war against pollution will be carried on throughout the nation rather than only in particular geographical areas." Federal emission standards "will preclude efforts on the part of States to compete with each other in trying to attract new plants and facilities without assuring adequate control." By making each state an air quality control region, by expanding and streamlining federal enforcement powers and procedures, much time would be saved for more direct and effective state abatement action.[15]

In the House debate on June 10, 1970, there was no real objection to giving the federal government authority to establish emission standards. After all, the bill did give states authority to set more stringent standards. Several speakers expressed great satisfaction with the federal standards. The new division of powers meant that standards could be revised with a minimum of administrative delay. With each state being an air quality control region, there would be a simplified and effective chain of command with "state governments directly in line from information to ultimate action."

Some wanted to expand federal power even further. Why not give the secretary of HEW authority to set national emission standards also for existing stationary sources? And why take the long road via federal air quality criteria? To this, the committee chairman, Mr. Staggers, answered that it would put "about everybody on the payroll of the United States." Instead of such an impractical solution, the committee bill allowed for considerations of local variation. If states wanted stronger standards to take care of particular problems in particular areas, the law should not prevent them by prescribing rigid federal standards.[16]

But why not give the states such power in the field of auto emission control? Why have a provision preempting states and local communities from setting their own, more stringent standards for auto emissions? With the National League of Cities pressing for local auto emission standards to take care of a pollution problem common to most metropolitan areas, why make an exception only for California? In arguing for his amendment to give all states the same right as California, Mr. Saylor of Pennsylvania pointed out that "the same conditions that exist in California exist in every metropolitan area of the United States." He wanted other states to have the same right as California in setting standards they deem necessary for the health and safety of their people.[17]

Faced with this drive for local experience and decentralization, the committee spokesmen pointed to the dreadful consequences that would occur if the House split the bill's "provisions in different ways and let the states go their own ways," as Mr. Staggers put it. The ranking Republican on the Committee, Mr. Springer, said that "you could have 15 or 20 states which could prohibit the use of an automobile on the highways of that state unless you had a motor that conformed with the standards of that state. You cannot be any more ridiculous than that." Mr. Rogers of Florida, chairman of the Subcommittee on Public Health and Welfare, concurred: " . . . you just cannot drive from one state to the other if we permit this type of thing without everybody paying a fine or else having his car pulled off the road." Another member of the House said that there was no need for "fifty different standards that are going to involve all of us in innumerable and immeasurable disputes."[18]

The Saylor amendment was defeated in a teller vote, and the committee bill sailed through with only minor changes. The debate in the House showed the existence of a majority in favor of considerably expanded federal powers to establish national pollution control standards. If Senator Muskie wanted to keep his position as the leading environmental policymaker, he would certainly have to reassess his position on the allocation of policy responsibilities.

National Air Quality Control
The report issued by the Senate Committee on Public Works on

September 17, 1970, fully confirmed that Muskie and the other committee members were prepared to let their preference for public health be accompanied by a strengthening of federal authority.

The tendency to give a wider meaning to the concept of "federal assistance and leadership" was evident from two features of the report: (1) increased federal authority to establish standards and guidelines for air quality and emission control and (2) increased possibilities for federal control and takeover of state activities concerning implementation and enforcement. Not only were Muskie and other committee members in favor of federal establishment of national air quality standards and goals, but they also favored broadened federal power to set standards for stationary sources. The secretary of HEW would be responsible for establishing national standards of performance for new stationary sources, as well as for hazardous pollutants. But the committee went beyond the Presidential and House bills by proposing federal standard-setting for "selected pollutants," such standards to apply also to existing sources.[19]

The committee also proposed to vest power in the federal government to control the quality, and even to take over the development, of state implementation plans. Whenever the federal government decided that state efforts to develop an implementation plan for the whole or part of that state were inadequate or inconsistent with the intent and time schedules of the act, or with federal standards, it could step in and take over the development of the plan. The federal government could even substitute its own plan for that of the state. The committee swept aside the old cumbersome enforcement procedures that required state consent to federal action. It gave the federal government authority to institute enforcement action whenever it found inadequate enforcement of state implementation plans or inadequate abatement of violations of federally established standards. The federal government would also have powerful means to bring states into line through grants for support of air pollution planning and control programs. Assistance would be given only to programs performing adequately, according to federal judgment.[20]

With auto emission standard-setting and the testing of vehicles and engines still a federal responsibility, what would be left for the states? The states would be free to establish more stringent standards than those approved by the secretary of HEW. The states would still be mainly responsible for the development and enforcement of implementation plans. But as we have seen, implementation plans had to be federally approved, and federal takeover was a strong possibility. States would set standards for existing stationary sources as part of the implementation plans, but only to the degree that this power was not otherwise vested in the federal government. States would at some future time be responsible for the certification and licensing of new stationary sources. The committee said this federal responsibility should be delegated to the states as soon as possible.[21]

In many ways, the committee foresaw an important role for the judicial system in the development of air pollution control policies. Of course, the bill contained provisions for challenging administrative decisions, such as denial of certificates, in court. There were also provisions for court trial of violation of standards. The bill provided for review of auto emissions and deadlines and administratively developed standards and guidelines. While the committee was unanimous in proposing judicial review for the latter category, it split on the review of auto emission deadlines. The majority favored judicial review, but Republican Senator Robert Dole argued for Congressional review. He said that a decision to suspend the 1975 deadlines for one year was an inherently political one and should be taken by Congress on the basis of recommendations from the secretary of HEW. The bill also provided for policy review within the administration, through the secretary of HEW, of the impact on clean air policy of agency proposals. It also provided for yearly reports to Congress on the progress of federal and state implementation of the Clean Air Act Amendments.[22]

In the words of Senator Cooper of Kentucky, ranking Republican on the committee, Muskie and the committee thus not only accepted the establishment of national air quality standards, but also moved "a long way toward national emission standards—a concept rejected by the committee in 1967 as logical for moving sources but not for stationary sources." Senator Cooper foresaw "a degree of federal control beyond anything I have supported in the past." As we previously concluded, this acceptance of increased federal authority was motivated by the emphasis on protection of public health as the main policy objective. Health protection requires the same levels of air quality throughout the nation; hence the need for a consistent, national system of standards and the necessity to increase federal standard-setting authority.

However, the committee stated, "the establishment of ambient air quality standards alone has little effect on air quality," so it emphasized that "the implementation plan is the principal component" of the control efforts necessary to achieve national standards within the time schedule laid out in the bill. Several consistency and efficiency arguments were used to support total federal authority to develop implementation plans and enforce pollution control regulations. Thus the committee recognized that the authority available under existing law had not "been adequate to move quickly to abate violation of standards." Furthermore, the provisions had not "been used to the fullest extent practicable."

Several arguments citing precedents spoke against such a choice.

The committee recognized that the primary responsibility for air pollution control had rested with the states and localities and it respected this principle. Thus, states, localities, and intermunicipal and interstate agencies should be free to adopt standards and plans to have a higher level of air quality than that approved by the federal government.

As we have seen, the committee's solution was to strike a compromise between two seemingly conflicting objectives. Effectiveness and consistency would be served through an elaborate system of federal checks, balances, and incentives, the function of which would be to secure an adequate quality and pace of state implementation plans and enforcement activities. Federal intrusion into state domains would take place only if the states failed to do what was expected of them. As a last resort, a state governor might go to court to get an extension of the time limit for achieving air quality standards.[23]

The allocation of power was debated on the Senate floor only in relation to auto emission standards. The committee bill contained proposals for specific numerical standards to be achieved before specified deadlines. If Congress thus sets the standards, should it not also decide on suspension of the deadlines for achieving the standards? As has been mentioned, the committee was split on this issue. The Dole amendment would have vested authority to make the suspension decision in the Congress, while an amendment offered by Republican Senator Gurney would have let responsibility rest with the secretary of HEW. The majority supported a proposal offered to the committee by Republican Senators Cooper and Baker, providing for judicial review of the suspension decision.

Both Cooper and Baker used the character of the issue and the need to provide for due process as arguments in support of their proposal. Cooper said that all administrative decisions should be reviewable in court, and did not see any reason why constitutional rights to due process and rights of appeal should not apply also to air pollution control decisions. Baker argued that a decision to suspend the deadline would be one of fact—of determining whether manufacturers could or could not reach the deadline by applying good faith efforts. The character of the issue and the imperative of due process made the 1975 Congress—surrounded by environmental and industrial pressure, and with the 1976 election around the corner—the least suitable place for reviewing the suspension decision.

In rebuttal, both Dole and Gurney used arguments of consistency and effectiveness. Congress had already made the policy—set auto emission standards—so it should also take responsibility for reviewing that policy. Indeed, Senator Dole said, the suspension decision was also a political one, involving delicate social, economic, technological, and public health problems. With the technological and health record developed by the secretary, Congress should be in a good position to make the final judgment. Furthermore, court review would be time-consuming and cause delays in implementing the act. Gurney argued that technological and health aspects were most important, and that because of the secretary's expertise on these matters, he should be the one to make the final decision.

Muskie rallied to Dole's support, saying that Congress should be responsible for both making and reviewing auto emission control policy. He also said the courts lacked the technical knowledge necessary to review suspension decisions. However, Muskie's support was not enough to save the Dole amendment. The majority seemed content with the statement made by Senator Randolph, Democratic chairman of the Public Works Committee, that "if we sustain the principle of court review . . . , Congress can still act whenever conditions seem to require it." The Senate voted against both the Dole and the Gurney proposals and in favor of judicial review.[24]

Protecting State Involvement
During the period in which the committee of conference was considering the bill, all functions of the secretary of HEW under the Clean Air Act were transferred to the administrator of the Environmental Protection Agency, pursuant to the president's reorganization Plan No. 3 of July 1970. In his message, President Nixon pointed out that the assignments of departmental responsibilities did not reflect the interrelatedness of the environmental system; crisscrossing responsibilities along both media and pollutant lines did not contribute to successful control of pollution. In organizational terms, this required "pulling together into one agency a variety of research, monitoring, standard-setting, and enforcement activities now scattered through several departments and agencies." The National Air Pollution Control Agency—subject to so much criticism from the House committee—would thus be moved from HEW to the EPA. This reorganization "would permit response to environmental problems . . . beyond the previous capability of our pollution control programs," said the president, and this would especially hold for the setting and enforcement of air quality and emission standards.[25]

Consequently, all references in the bill to the secretary of HEW were changed to the EPA administrator to reflect this reorganization at the federal level. But this was not the only change in power allocation made by the conference committee. As one member of the Senate staff said: " . . . the House has a stronger view to protecting the states than we do. We are sensitive to that, but the House is somewhat stronger. They think about state involvement before federal involvement."[26]

Thus the conference committee stripped the administrator of the power given to him by the Senate to withhold funds from any local, state or interstate agency judged by him as inadequately staffed to carry out its responsibilities under the act. The language of the Senate bill was changed to make it clear that the states would have the primary responsibility for assuring air quality within their entire geographic area. Chapter 3 showed that the conference bill distinguished between primary and secondary air quality standards. States were allowed to determine a reasonable time for reaching the secondary standard, and the possibilities of getting extensions of the three-year period for reaching primary standards were somewhat increased. The conference bill dropped the Senate provisions for federal preconstruction review and certification of new emission sources. The only thing left was a requirement that state implementation plans must contain a provision for state review, prior to construction, of the location of such new sources. With respect to enforcement, the Senate bill made no distinction between violations of state implementation plans and federal emission standards. However, the conference substitute provided that the EPA must give the state a thirty-day notice before acting on violations of implementation plans, whereas the EPA could act directly on violations of federal emission standards.[27]

In the Senate debate, Muskie said he had been interested all along not only in increasing federal presence and backup authority but also in preserving "local option" features, to allow the states to respond to their particular air pollution problems in the most efficient way. He believed that the bill would give state and local authorities sufficient latitude in selecting ways to prevent and control air pollution. But the changes proposed by the conference committee and adopted by the two houses in December 1970 were not drastic. In the words of Mr. Springer, Republican House conferee, they were "matters of language and emphasis." Federal authority to establish auto emission standards and ambient air quality standards, as well as national emission standards for new sources and hazardous pollutants, was retained. The provisions for federal EPA control and approval of state implementation plans was retained and would serve as a means to prevent inconsistencies in the development of policy implementation. The main thrust still was to increase federal powers. As Senator Cooper had already said in September, there was "a degree of federal control beyond anything . . . in the past . . . but one we believe is *necessary as we begin to deal with the air pollution as a national problem.*"[28]

Summary

The Swedish choice was to centralize almost all policy authority at the national level in the NEPB, the central environmental agency. Policy interpretation (i.e., the development of standards and guidelines) was made the principal responsibility of the NEPB. It is true that performance control and enforcement was made a primary responsibility of the state regional boards, but the NEPB would be omnipresent, thanks to its coordinating function. On two points there were differing opinions: (1) who should make individual licensing decisions? and (2) who should be responsible for future policy planning and review? The licensing authority was finally vested in the FBEP, a quasi-judicial central administrative agency. However, the NEPB was given wide influence and authority through the exemption procedure. A minority in the Riksdag wanted to keep the responsibility for policy planning and review within the legislative realm. However, the Social-Democratic majority seemed quite prepared to vest policymaking and reviewing authority in the central environmental agency. It is also important to note the very limited role given to the judiciary.

The Social-Democratic majority wanted this strong concentration of policy responsibilities in the central agency to secure efficient and consistent implementation of environmental policy. By concentrating so many functions in one agency, there would be enough expertise and experience to guarantee consistency and efficiency. The character of most pollution control issues—trade-offs with large-scale social, economic, and technological ramifications—also warranted centralization within the administrative realm, and spoke against allocating responsibilities in the judiciary. There were, however, limits to this centralization. The total concentration of power envisaged in the NEPB's proposal for administrative licensing lost out, thanks to the resistance from opposition parties and future licensees, who argued for a separate franchise board to secure due process and the rule of law.

The United States choice was to increase substantially the federal role in standard-setting and enforcement, but also to keep the states and the judiciary involved in a pattern of checks and balances. In the future, standard-setting would be a federal concern, except for existing stationary sources. State implementation and enforcement would become subjected to increased federal authority to check the quality of state implementation plans and state enforcement performance. The federal EPA would have the authority to take over state implementation and enforcement whenever this would be necessary to achieve national standards. The traditional state authority to establish standards higher than the federal ones was retained; however, California was the only state allowed to do so with respect to auto emissions. The judiciary would continue to play an important role in the policy process through the provisions for judicial review of decisions to suspend auto emission deadlines and other administrative regulations and orders. In fact, the role of the judiciary could be expected to increase because of the provisions for citizen involvement in the enforcement of policy.

Spokesmen for increased federal leadership and assistance pointed to the lessons of existing policy. The fragmented approach of 1963 and 1967 simply did not work; inconsistency and inefficiency had led to the occurrence of pollution havens. To protect public health and to have a consistent and efficient national clean air policy, centralization of authority to the federal level was necessary. At the same time, there was a strong demand for continued state involvement and for due regard for state initiatives. Local variations in air pollution made local experience indispensable to a successful development of air pollution control. But the end result was increased federal authority; consistency and efficiency constituted the narrow limits within which local experience and continuity would be allowed to play a role. Compared to Sweden, there was a stronger and more unified concern for constitutionality and due process. For these reasons, judicial review was preferred to Congressional review in the case of suspension decisions. The majority of Congressmen supported Baker's argument that such review was judicial in character, as well as Cooper's plea for due process under the law.

Thus, both countries expanded the role of central government in the policy process. The difference is that Swedish policymakers could centralize competence without much hindrance from constitutional law or institutional precedents, while United States policymakers had to observe constitutional rules and institutional precedents when expanding federal responsibility for a continued, albeit thoroughly renewed, clean air policy.

CHAPTER 5

Public Participation in the Implementation of Policy

Concepts and Arguments

In this chapter, we will deal exclusively with modes for public participation in the implementation of public policy. Arrangements for public participation in the formulation of policies to be adopted by legislatures—policymaking—fall outside of our analysis.

First, let us clarify what we think the concept should connote. It seems necessary to make a distinction between those individuals and groups explicitly mentioned as "target groups" or "clients," and all other individuals and groups. The participation of the first category comes naturally, as a result of the status as regulated or client. What is of particular interest here is to what extent, and in what ways, other individuals and groups are allowed to enter the process of implementation. Within this public category, one may distinguish between those allowed to participate in their special capacity as "affected parties" or "concerned interests" and those given access in their general capacity as "citizens." The first category may be given access to defend or promote their individual interests or concerns, while the second category may be allowed to participate to promote the public interest.

Second, a distinction must be made between the different arenas or channels open to the public as well as among matters of policy substance open to participation and possible influence. The main arenas are administrative agencies and judicial courts. The policy substance may concern either or both of these categories. It may concern the process of interpreting the legislation to establish general rules and guidelines for implementation, and it may concern implementation in individual cases. Participation in a particular case may either precede or follow the implementative decision. The affected party or citizen may thus be given an opportunity to influence the content of the decision, or he may be limited to seeking changes or redress afterwards. Individuals or group representatives may act as members of agency boards or consultative bodies, and they may participate because the implementative rules give them a special position in the decision-making process as such.

The policymaker wanting to promote public participation has a number of techniques to choose from. Public hearings can be made compulsory in the process of determining implementation guidelines. Hearings may also be made part of the process in individual cases, often coupled to requirements for public site inspections. Consultative procedures to allow public involvement, as well as representation on advisory boards, can also be used to increase or ease participation. Usually, possibilities for administrative or judicial complaint are written into a policy. Affected parties or citizens may be allowed to institute proceedings to speed up regulatory action or to seek compensation for damage.[1]

The Case for Participation

When it comes to arguments for increased public participation in the implementing stages of the policy process, two main categories can be listed (see fig. 9). According to one line of argument, participation is a goal in itself. Increasing citizen participation is simply a matter of sound and desirable policy to be promoted in as many ways as possible. The idea of an equal right for everybody to be involved in decisions concerning the general public is frequently voiced. Another line of reasoning sees participation simply as a matter of strategy—a way of accomplishing other overt or covert objectives. Participation may often be used as a means for gaining legislative and political support and consent. Most politicians and bureaucrats in democratic systems prefer to act from positions of strong public support. Thus, planning for public participation to gain such support is a natural strategy, and increased participation may be a means of getting popular consent for unpopular policy ideas.

Some argue for public participation as a means of increasing

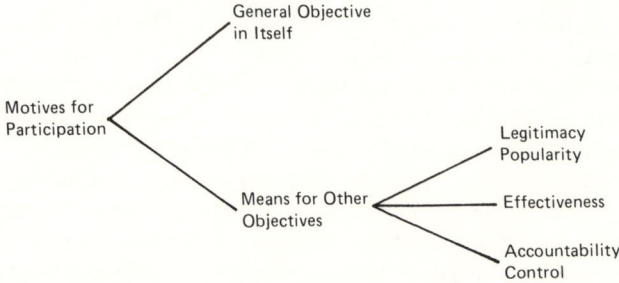

Fig. 9. A classification of motives for allowing public participation in the implementation of public policy

administrative efficiency and effectiveness. Participation, they hope, will bring improvements in information, communication, and understanding of the views of different parties. A related argument concerns avoiding efficiency-decreasing conflicts by arrangements that bring affected parties together in agencies, boards, and councils.

A third argument concerns the accountability of agencies. The control of administrative decision makers is likely to be reinforced if the process is open to public participation and scrutiny. Participation and ensuing publicity provide pressures that help ensure that administrators and officials will act fairly and follow required procedures in all cases.[2]

How does all this fit in with the implementation of air pollution control policy? In one sense, every citizen is involved, since everyone must breathe. Large-scale air pollution affects people at long distances from the source. On the other hand, it is clear that people living close to an emitting plant are more affected than those living far away. On the grounds of pollution proper, one would expect to find similar arrangements and arguments concerning participation in air pollution control in most industrialized, urbanized, and motorized nations. But politicians' choices and arguments also reflect their considerations of consistency with cultural norms, institutional arrangements, and strategic circumstances. At issue then, is whether—and how—such considerations play a significant role in the choice of modes and levels of public involvement in policy implementation.

Sweden Argues for Limits to Participation

When the Swedish Riksdag discussed the Environment Protection Act in 1969, several measures had already been taken to institutionalize interest group participation in environmental policy implementation. The NEPB was established in 1967. At that time the minister of agriculture proposed that the NEPB be governed by a board of directors, consisting of a director-general and six other members. As far as possible, the six should represent interests especially concerned by, or having knowledge and experience especially relevant to, the NEPB's future decisions.

But six people could not represent all aspects of the NEPB area of competence. Therefore, the minister wanted to establish special advisory councils to assist the NEPB and the director-general on important matters of policy implementation and reassessment. The minister saw the proposed Air Quality, Water Quality, and Nature Conservancy Advisory Councils as important "impulse givers" within their policy areas. They would increase the agency's capacity for solving pollution problems. Efficiency and effectiveness were clearly the main aim of the proposal. The Center party used a somewhat different line of argument. It was essential that the new policy have the broadest possible popular support and legitimacy. Thus, the layman's opinion should be heard in the councils through representation from business, trade unions, and the association of local communities. In effect, the Air Quality Council's membership has been very much in line with what the Center party proposed.[3]

As shown in earlier chapters, Sweden's approach to air pollution control is basically one of individual source control. The 1969 act provides only very general and flexible rules to guide NEPB and FBEP decisions in individual cases. A system of emission guidelines was seen as necessary to help the agencies interpret the law in a coherent fashion. In theory, one could argue for such a system of implementation guidelines along at least two lines. A system of guidelines would define an agency's ability to exert discretionary power, and thus increase control over administrative performance. In addition, a system of guidelines would also improve the agency's capacity and performance. The guidelines would function as the agency's rules of thumb and as information to polluters concerning their responsibilities. The question is whether proponents of a system of implementation guidelines also used such arguments to support proposals for some form of public participation in preparing such a system of guidelines.

Increased efficiency and effectiveness was clearly the main motive throughout the Swedish debate on emission guidelines. The Royal Commission on Pollution and Other Nuisances stated in 1966 that guidelines would standardize the franchise process, speed up decisions, and make them more predictable. Polluters would adjust their precautionary measures and permit applications to the guidelines, which would make implementation even more effective. The commission found it natural that guidelines "should be established by the responsible agency after negotiations with the concerned industrial branch organizations and others with an interest in the matter." Participation of these parties in the preparation and implementation of guidelines would "form the basis for a close and efficient cooperation between the responsible agency and the industry." The guidelines should be available to the public and subjected to public debate.[4]

Several months before the commission's report was published, the then existing National Air Pollution Control Council had approached industrial branch organizations, inviting them to participate jointly to work out emission guidelines for different industrial branches. The council wanted a smooth and efficient introduction of the new environmental legislation, and felt that technical and economic background data for a system of emission guidelines could best be provided through a joint agency-industry effort. There was no discussion of broader participation in the groups. In the ten groups established up to March 1969, 70 percent of the members represented industrial interests and 20 percent scientific and technological expertise, while the remainder came from within the NEPB. Consumer and citizen groups were not represented. Industrial and NEPB representatives afterwards agreed that their participation in the joint working groups served as a vehicle for improving the efficiency of policy implementation.[5]

Thus there was already a system of formal and informal interest-group participation in policy implementation when the Riksdag discussed the new environmental legislation in May 1969. Since the system had developed outside the formal framework of the new act, the cabinet proposal and the parliamentary discussion had to address the questions of who should participate, when and where participatory rights could be exercised, and what substantial matters the right of participation should concern.

The question of who should participate in implementation was never really discussed in terms of choosing between a citizen or an affected-party approach. At issue was, rather, how widely defined the latter approach should be. The 1966 Royal Commission had recommended a very narrow definition. Only those with ownership or any other legally valid title to property suffering damage or nuisance from a particular polluting source under administrative or judicial scrutiny should be allowed to enter the implementation process.

The minister of justice, Mr. Kling, argued that not only property owners were disturbed by air-polluting activities. Large-scale pollution affected such large numbers of people that controlling it was in most cases in the public interest, although each and every disturbance did not constitute a valid reason to initiate or take part in implementative actions. For example, encroachments on the public's customary right to common access to the Swedish countryside was not, in the minister's view, a valid reason. Indeed, this argument from the minister showed that a citizen approach to participation was out of the question. *Individuals* could not call for enforcement of the act's provisions to protect the *public* interest. They could act only to protect their own individual interests. The minister finally argued that legal precedent established in the Water Courts could be used to determine who should be considered an affected party with a right to participate.

The cabinet's Legislative Advisory Council reacted acerbically to the minister's arguments. Legal tradition defined the number of affected parties exactly the way the 1966 Royal Commission had recommended. This would—unacceptably—favor property owners. Instead, the council insisted, all who suffered damage or were subjected to a nuisance from polluting activities should be viewed as affected parties with a right to defend their interests in the implementation process. The only condition was that there must be a cause-effect relationship between the polluting activity and the damage. The minister accepted the council's argument. To appear as an affected party, it would be enough to show that one had suffered some discernible damage from a particular polluting activity. No property title would be necessary.

The minister's main argument for limiting public participation was that widespread public involvement in policy implementation simply was not necessary. Policy measures in the past few years—establishment of the NEPB in 1967 and strengthening environmental departments within the twenty-four state regional boards—had increased administrative capacity to control pollution. The new act established a system of franchise and advance notice that would cover almost every individual case of pollution. This meant that administrative action would serve to satisfy not only the public interest in a clean and healthy environment but also the individual interest in avoiding damage and eliminating nuisance. The functions earlier performed by affected parties could now be much more effectively carried out by administrative agencies. In the minister's own words: "The more we intensify the governmental efforts to protect the environment, the more satisfied the individual's interest in an undisturbed neighborhood will be!"[6]

This general view of the relationship between individuals and public agencies no doubt guided the minister in his proposals concerning the three processes envisaged for controlling individual sources of air pollution: franchise or permit proceedings in the FBEP, exemption decisions in the NEPB, and court proceedings in the Real Estate Courts (see fig. 10). The latter could concern both the permissibility of the polluting activity and the question of compensation to individuals suffering damage from the activity.

The polluter had two choices under this system. First, he could go directly to the FBEP and apply for a permit. The permit would be legally binding for ten years. Second, he could ask the NEPB for exemption from the duty to apply for a permit. In both cases, the conditions and precautionary measures necessary to continue the polluting activity would be spelled out. The exemption could be revoked at any time by the NEPB.

The minister characterized the exemption proceedings as "free negotiations among the NEPB, the applicant, and other interested parties." The NEPB would insert a public notice in one or more local newspapers or otherwise inform affected parties about the polluting activity for which an exemption was being sought. Site inspections and hearings with affected parties would be held only if the NEPB decided that this was necessary for its own decision making. The concept of locality indicated that affected parties were only those living in the

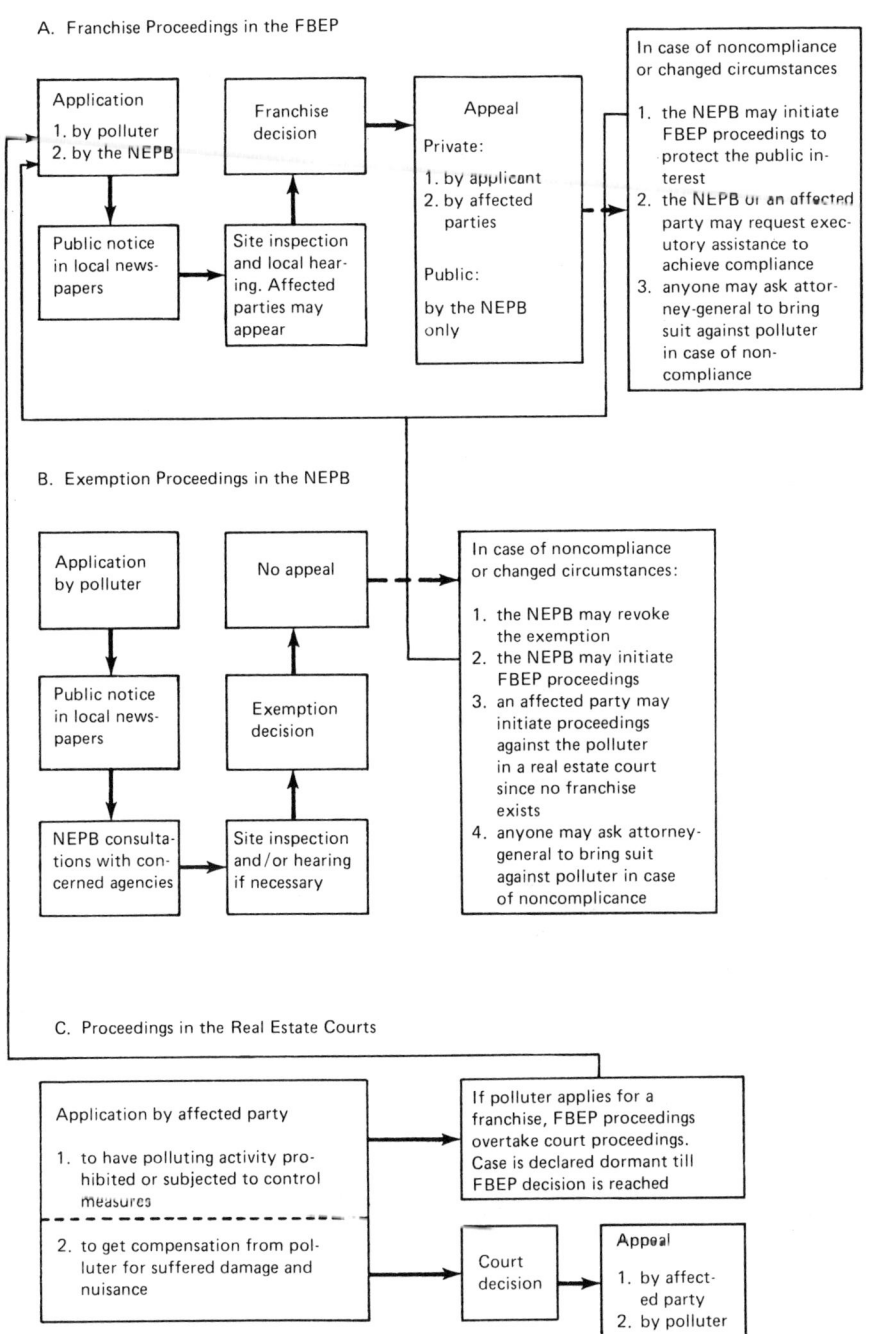

Fig. 10. Procedures and participatory channels in Swedish pollution control

neighborhood and possibly exposed to nuisance or suffering damage from the activity. Since the decisions resulting from these free negotiations were not legally binding, there would be no right to appeal. The polluter could always apply to the FBEP for a permit, and the affected party could resort to suing the polluter in a real estate court to have the polluting activity prohibited or circumscribed.

But should not the affected party be able to take the exemption decision to the franchise board? No, said the minister. The NEPB and the FBEP were to take public interests into account in their decisions. Since both bodies were now substantially strengthened, their activity would increasingly protect and benefit individual interests. Lingering differences of opinion between polluters and affected parties should be dealt with in the real estate courts. Nor should the administrative agencies, with which the NEPB was to consult in exemption cases, have the right to take the issue to the FBEP. Why? The NEPB had already taken all aspects of the public interest into account in its decision!

Thus only the polluter and the NEPB had the right to bring the issue to the franchise board. Only the NEPB would have the right to ask for franchise proceedings to protect the public interest in a case of air pollution. The proceedings would, as a rule, include site inspections and public hearings in the locality. Affected parties would thus always have the opportunity to state their case, and the FBEP was to take individual as well as public interests into account in its permit decision.

But who should be able to appeal a permit decision? According to the minister, it was "only natural" for the applicant and the NEPB to have such a right. It was much more doubtful whether individual affected parties should be allowed to appeal franchise decisions. To Minister Kling, since such parties would benefit from the strengthening of the public agencies, private appeal seemed unnecessary. In the end, however, the minister came out in favor of allowing private affected parties to appeal permit decisions, but with such expert and comprehensive procedures in the franchise board, the minister foresaw very limited effects of such a right. Only in exceptional cases would a private appeal provide opportunities for revoking or changing a permit decision by the FBEP.

Still, the affected party would have the opportunity to bring a suit in the real estate court. In cases where the polluter had no permit from the franchise board, the affected party could bring suit to have the polluting activity prohibited or subjected to precautionary measures. The polluter could then take the case to the FBEP by asking for a franchise. The court must then declare the case dormant till the FBEP reached its decision on the franchise issue. If a permit was granted, the real estate court trial could cover only the issue of compensation. As was shown in the discussion of who should be able to participate, only those being able to show a cause-effect relationship between the polluting activity and objectively discernible damage would be acknowledged by the court as legitimate affected parties.[7]

The cabinet's Legal Advisory Council was not content with the proposal with regard to an affected party's opportunity to defend his rights. The council seemingly accepted the public-private distinction proposed by the minister, the principle that the environmental agencies should represent the public interest while private citizens could only speak for their individual interests on environmental issues. But the council did not accept the minister's argument that strengthening the environmental agencies automatically benefited individual affected parties. The council felt that all such parties had an interest in influencing decisions at the earliest possible stage (i.e., during the franchise proceedings), but the minister's proposal included obstacles to such an early appearance. In the existing Water Courts, legal practice all but guaranteed that interested parties would be compensated for their costs of appearing and bringing forth evidence. No such possibility existed under the proposed act. Consequently, the proposal made the situation of affected parties worse than it had been under the Water Act. In addition, the NEPB in many cases might come out in favor of solutions going against the interests of individual affected parties. The council strongly recommended changing the proposal so that such parties would get compensation for their costs of appearing at FBEP hearings.

The minister expressed no sympathies for the council's arguments. Again, he showed his faith in the agencies' objectivity and competence. The FBEP had a duty to bring forth a comprehensive and objective investigation, and to take into account both public and private interests. In most cases, the minister believed, these interests would indeed coincide, and affected parties would thus benefit from agency activities and decisions.[8]

The opposition parties supported the viewpoints of the Legal Advisory Council. The Center party and the Conservatives wanted better chances for individual affected parties to state their cases and defend their interests. The cabinet proposal undermined these possibilities, since it did not provide compensation to affected parties for the costs of bringing evidence and using legal counsel. Those affected would find themselves in an awkward position because pollution issues are generally very complicated, both legally and technically. These difficulties were compounded by the fact that small groups and individuals might often find themselves in a position adversary to the NEPB, "the spokesman for the public interest." Both parties proposed a change that would give affected parties a right to compensation for the costs of participating in the process. The Center party also proposed that not only the polluter but also the affected party should have the right to initiate proceedings in the franchise board.[9]

The Liberal party went even further than discussing a more equal position for the affected party as compared to the polluter. The Liberals said that the cabinet proposal, with its neglect of, and even contempt for, the rights of affected parties, revealed "a seriously biased approach to the whole environmental problem." Liberals contended that consumer and neighborhood interests suffered most from polluting activities yet, in the present cabinet proposal, were the ones most separated from participation in pollution abatement. The effect of this separation could only be detrimental to the public interest in a better environment. Crucial information would not be forthcoming to the agencies, and serious pollution cases might receive less and later attention than would otherwise have been the case. Affected parties should be entitled to have all matters of compensation settled in advance by the franchise board, the Liberals argued, and local communities should be given a right to appeal franchise decisions affecting a public interest in the local community.[10]

The Riksdag debate repeated all of the earlier arguments. A Social-Democratic majority spokesman reiterated that all affected parties would benefit from the new act, which strengthened governmental involvement in pollution control. The most important thing was that the act required the FBEP to provide a thorough investigation in every case. Therefore, "there would rarely be any need for special investigations by individual 'affected parties.' Consequently, there is no need for special compensation rules." Countering these arguments, a Liberal party representative exclaimed: "The individual, whose eyes, ears, and nose react [to pollution], has no reason whatsoever to rely on Big Brother in Stockholm, who is supposed to defend his interests, . . . but the committee calms everyone down by saying that environmental interests are best taken care of by the NEPB. *Big Brother thinks of everything!*" All efforts to increase the possibilities for participation were defeated, and the minister's proposals became law.[11]

The United States Provides for More Participation

It is easily forgotten that the Clean Air Act Amendments of 1970 were amendments to existing legislation. The laws passed in 1963 and 1967 contained several provisions for public participation, such as conferences, public hearings, and judicial review of agency decisions. The 1970 amendments proposed in the administration and House bills concerned new pollution control policy concepts, such as national ambient air quality standards, performance standards for new stationary sources, and standards for hazardous pollutants. Also new was the idea of tying state implementation plans to the national air quality standards. Both bills preserved earlier provisions for public participation, and also contained provisions for such participation in the interpretation and implementation of the new policy concepts.

Among the features adopted from earlier legislation, two very much resemble Swedish arrangements for interest representation in policy development and implementation. Within the EPA, there was to be an Air Quality Advisory Board, chaired by the EPA administrator and having fifteen members, who represented government agencies at the state, interstate, and local levels. Public and private interests—polluters as well as those affected by pollution—and others demonstrating an active interest in the field of air pollution prevention and control were also represented. The board was to advise and consult with the EPA administrator on policy matters and to make recommendations to the president. From time to time, the administrator could also establish advisory committees. These expert panels are to assist on special air pollution problems from the viewpoints of health, welfare, economics, and technology.

Both bills provided for public hearings before the adoption of ambient standards, state implementation plans, and standards for hazardous pollutants. Interested persons would have a period to give written comments on EPA proposals concerning new source standards and ambient standards. However, these provisions received very little attention during the House debate in June 1970. No arguments concerning the merits or possible drawbacks of participation were presented.[12]

It was the Senate, especially the Muskie subcommittee, that presented some innovative concepts for public participation. It was also the Senate that presented the most elaborated and coherent arguments for increased public participation in pollution control. In one stroke, the Muskie subcommittee and the full Committee on Public Works increased the possibilities for participation in the development of auto emission standards, broadened the scope of judicial review of administrative decisions, and presented the truly innovative concept of "citizen suits" as a means of policy implementation. (For a full picture of the provisions for public participation in the legislation, as finally adopted, see table 3.)

It is quite instructive to compare the Senate arguments with those presented in the Swedish debate on public participation. Nowhere does one find any preoccupation with limiting either the number of legitimate participants or the scope of their right to participate. On the contrary, the committee report and the debate are filled with statements concerning the public's right to participate. The citizen, not the affected party, is seen as an important participant, not only in the development

TABLE 3. **An Outline of Possibilities of Public Participation in the Implementation of U.S. Air Pollution Control Policy**

Administrative Agencies*
Implementation rules and guidelines
Representation
1. President's air quality advisory board (15 members)
2. EPA advisory committees. Expert panels to be established at EPA Administrator's discretion.

Statutory right to give written comments
1. Primary and secondary ambient air quality standards. Ninety-day period provided for all interested to give comments on proposed standards.
2. Performance standards for new stationary sources (same provision as above).

Public hearings
1. Adoption of state implementation plan. State adoption after reasonable notice and public hearing.
2. Revisions of state implementation plans (same provision as above).
3. EPA takeover of state implementation plan. Public hearing, if state failed to hold one.
4. National emission standards for hazardous pollutants. Public hearings within 30 days of proposal of such standards.
5. Suspension of auto emission deadlines. Public hearing must be held.
6. Aircraft emission standards. Hearing in regions most heavily affected.

Application to individual cases
Public hearings
1. Interstate pollution. EPA may call a conference on potential problems in a particular area; all interested can attend. If no abatement action is forthcoming as a result of the conference, EPA is to call a public hearing on site of pollution.
2. Mobile sources not conforming to auto emission standards. Public hearings if manufacturer disagrees with EPA decision on remedies manufacturer and interested persons may attend.
3. Fuel regulation. If EPA proposes regulations and affected business so requests, a hearing should be held.

Judicial Court
Implementation rules and guidelines
Judicial review
1. State implementation plans. All determinations subject to judicial review. Petition may be filed by any interested person within 30 days of decision. Courts of appeal have jurisdiction to set aside determination appealed as a whole or in part.
2. EPA decisions on ambient air quality standards; performance standards for new stationary sources; national emission standards for hazardous pollutants, auto emission standards and emission standards and suspension of deadlines; regulation of fuels; aircraft emission standards. Subject to judicial review. Petition may be filed by any interested person.

TABLE 3. (Continued)

Application to individual cases
Citizen Suits—any person may commence a civil action on his own behalf against:
1. a person, including governmental agencies, who is alleged to be in violation of an emission standard or limitation or an EPA or State order issued under the Act; or
2. against the EPA administrator, where he allegedly fails to perform his mandatory functions.

The courts have jurisdiction "without regard to . . . the citizenship of the parties" to enforce the standards, limitations, orders, and to order the adminstrator to perform his functions, as the case might be. A 60-day notice of the intent to file suit must be placed with the state or federal agency, to provide time for abatement action.

---

*Open Files. Polluter and agency documents on stationary sources, implementation plans, new sources, hazardous pollutants, manufacturer records and reports on mobile sources publicly available if a trade or product secret is not divulged. All results of EPA motor vehicle testing publicly available.

of guidelines for implementation but also in the enforcement of such guidelines in specific cases. Nowhere does one find arguments to the effect that an increasingly efficient administration's activities will make public participation unnecessary. Citizen involvement is seen as a valuable complement to administrative action.

This emphasis on the rights of the public is clearly seen in the arguments concerning open files. Arguments contended that participation in the development of state implementation plans could only be meaningful if there is reasonable notice and full disclosure of information prior to public hearings. All information relevant to pollution from new stationary sources, from hazardous pollutants, and from auto exhausts should be available to the public. Information other than emission data should be considered confidential only after owners and operators showed that such information would divulge trade secrets. Public knowledge of emissions should override the private interest in proprietary information: ". . . it is not in the public interest for data relating to the quantity and quality of the emissions to be considered confidential. The public has a right to know who is polluting the atmosphere and in what amounts."[13]

But what should the citizen be able to do with this information? Clearly, the committee did not have the Swedish system in mind. Since the beginning of the nineteenth century, Swedish citizens have had constitutionally guaranteed access to governmental and agency files. But as we have seen, information gathered as a citizen is of no use to the Swede if he cannot prove his status as an affected party in the particular pollution case. Not so in the United States Senate committee proposal. The committee intended that Americans should be able to participate and influence policy decisions as citizens, and that such participation was equivalent to performing a public service.

Making petitions for judicial review of administrative decisions would be one way for the citizen to use available information to influence policy implementation. The committee referred to the courts' granting review to those seeking to protect the public interest in the proper administration of a regulatory system enacted for their benefit. The uncertainty in the existing law with respect to judicial review was undesirable, and the committee therefore included special provisions for such review. Any person or class of persons believing that a promulgation of ambient air quality standards, emission standards, implementation plan, and certain other administrative decisions should be modified in any way, was given the right to file a petition for judicial review in the United States Court of Appeals. Petitions had to be filed within thirty days of the administrative decision. Thereafter, review would only be granted whenever significant new information bearing on the decision had become available. Evidently, open files would become the vehicle through which interested citizens could judge the importance of new information and proceed to control administrative performance through petitions for judicial review.[14]

Control and efficiency were clearly the main arguments used in favor of the innovative concept of "citizen suits." While public hearings and judicial review would serve as means for citizens to influence and challenge the development of implementative rules and guidelines, citizen suits would concern the application of and compliance with these guidelines in individual cases. The committee proposed that civil action could be brought in district courts to seek the enforcement of compliance timetables, emission and performance standards, and other requirements and prohibitions established under the act. Any citizen or group of citizens could bring such action against any polluter allegedly violating such requirements, as well as against the EPA administrator if he allegedly failed to perform duties established under the act.

The committee argued that "authorizing citizens to bring suit for violations of standards should motivate governmental agencies charged with the responsibility to bring enforcement and abatement proceedings. . . . To further encourage and provide for agency enforcement, the committee has added the requirement that prior to filing a petition with the court, a citizen or group of citizens would first have to serve notice of intent to file such action on the federal and state air pollution control agency and the alleged polluter." In such a way, agencies would have a period of time to commence enforcement action. If the court then found that such administrative action was adequate, the case could be dismissed.

But could not citizen suits lead to inconsistency and inefficiency? The committee found this unlikely to happen. In the committee's view, the courts would try the issue of compliance with standards, and not establish any new standards. The suits and court decisions would include no provision for recovery of damage. Frivolous and harassing action would be rare; the provision that courts may award costs of litigation would serve as a deterrent in this respect. However, the committee stated, the courts "should recognize that in bringing legitimate actions under this section citizens would be performing a public service." In such cases, the courts should award costs of litigation to such party.[15]

Immediately upon publication of the Muskie subcommittee's proposal in late August 1970, pressure was brought to bear on the full committee to delete the citizen suits provision. Industrial comments challenged both the control and the efficiency arguments underlying the proposal, arguing that there were no grounds for the assumption that government agencies would fail to perform their duties and, consequently, no need for citizen suits to enforce policy requirements. Furthermore, the proposal made efficient and certain administration virtually impossible. The provision was "simply costly, contentious, and confusing—but not very effective," and should be dropped.[16]

But the Committee on Public Works held out, and kept the provision. Managing the bill on the Senate floor, Muskie could state that the bill "extended the concept of public participation to the enforcement process. The citizen suits authorized in this legislation would apply important pressure . . . they should be an effective tool." Senator Hruska said that citizen suits should not be considered a substitute for the enforcement efforts of the responsible agencies. Rather, he stated, they would complement and encourage the abatement activities of governmental agencies.

The bill's strongest critic, Senator Griffin, said the proposal would overload the judicial system. Frivolous action would indeed be the rule, he argued, because of the provision for awarding litigation costs; environmentalists would sue, hoping that the polluter would have to pay. Agencies would be tied up in endless suits, which would decrease administrative efficiency. Furthermore, arguments for increased citizen control and involvement were based on the erroneous assumption that the EPA and other agencies would not perform their duties. In short, Griffin said, the proposal "would retard progress, taxing the time, resources, and manpower of the agency."[17]

In rebuttal, Senator Muskie stated that the section did not presume any lack of good will or dedication on the part of administrative agencies, but the implementation of national ambient air quality standards would create potentially enormous enforcement problems. It would be too much to presume that these enforcement agencies, "however well staffed and well intentioned," would be able to take care of all potential violations. Citizens would therefore "be a useful instrument for detecting violations and bringing them to the attention of the enforcement agencies and the courts alike." Through advance notice to the agencies, the citizen might trigger enforcement action. Muskie assumed that most people would be satisfied with having done so, and would thus not go on to initiate court proceedings and overload the judicial system. Muskie's arguments for increased participation were indeed quite contrary to those used by the Swedish minister to prove that such participation would not be necessary. Said Muskie: "Citizens in bringing such action are performing a public service. The limited resources of many state enforcement agencies, bearing the first line of responsibility under this bill, will be fully extended. This provision, requiring thirty days notice to state and federal agencies, in which they may initiate abatement proceedings, will allow many violations to come to their attention which might otherwise escape notice." Furthermore, the right to clean air should not be limited to what the courts could comfortably handle and by denying citizens access to the courts; " . . . in a society of government of and by the people we foreclose participation by citizens at our peril."[18]

Some controversy arose around the question of whether or not the citizen suits provision also provided for class action. The committee in its report explicitly stated that it did not. Citizens would have the right to sue a polluter or the federal administrator for an injunction, but were not granted a right to sue for damages. However, the section did retain all rights that citizens might have under state and federal law, including class actions and traditional common law actions, to subsequent suits for damages. Furthermore, the committee at the same time talked about "actions instituted by any citizen or class of citizens." This language, much more than the qualifying "to enforce or require enforcement of certain provisions of the Clean Air Act," caught the attention of certain senators.[19]

These senators referred to the provision as "authorizing certain class actions." The call for class actions was in the language of the bill: " . . . regardless of the amount in controversy, or the citizenship of the parties . . . I think they are discussing and indicating a class action." Senator Muskie again and again stated that this was not the case. Citizen access to courts was much more limited than that contemplated in the pending consumer class action legislation. It was limited to citizens acting on their own behalf to seek abatement of violations of standards established administratively under the Clean Air Act Amendments. Muskie declared, "In other words, *the idea is to use citizens to trigger the enforcement mechanism.*"[20]

The bill passed. Since there were differences between the House and

Senate versions, meeting in conference was necessary. Amidst conference proceedings, the secretary of HEW sent a letter to the senior Senate majority conferee, asking for several modifications in the Senate and House versions. With respect to citizen suits, the secretary stated that they could contribute to an effective enforcement of pollution control measures. However, citizen suits against the administrator to force him to take enforcement action in a particular case must be rejected for reasons of effectiveness. It "would have the unintended result of reducing the overall effectiveness of our air pollution control efforts by distorting enforcement priorities that are essential to an effective national control strategy." He therefore recommended deleting the provision in the Senate bill authorizing citizen suits against the administrator.[21]

The conference committee was not prepared to go that far in changing the bill. The compromise bill provided for citizen suits to be commenced by "any person on his own behalf" against any person allegedly violating any standard, limitation, or order issued under the act. Suits against the administrator were limited to alleged failure to perform mandatory functions. The compromise bill extended from thirty to sixty days the time period required between prior notice to violators, the administrator, and the state, and commencement of court action. Suits against stationary source violations must be brought in the district where the source is located. The bill explicitly prohibited citizen suits in cases where the administrator or the state is diligently prosecuting civil action to enforce requirements made under the act.

In the House debate on the conference report, the provisions for public participation received only passing remarks. Evidently, the modifications made in the conference bill were sufficient to still potential criticism against House conferees for accepting a concept that had never been contemplated by the House. In the Senate, a colloquy between Senators Eagleton and Muskie touched upon the issue of public participation. Eagleton assumed that the provisions for public participation were likely to result in higher quality and better air pollution control programs than would be the case if there were less opportunity for citizen participation. Muskie said that this was the thrust of the new legislation. Participation by citizens would be a key element in the successful prosecution of air pollution goals. He and the Senate committee "felt it would be impossible to do the total job of air pollution cleanup relying wholly upon the federal bureaucracy."[22]

Summary

The Swedish choice was clearly geared to limiting the number of participants in the implementation process. Legitimate participants would be those affected parties able to prove a cause-effect relationship between a particular case of pollution and the damage suffered. Nowhere among majority politicians does one find a willingness to allow citizens to participate. The choice was also aimed at limiting the *scope* of participation. The affected party could participate only where his or her private interest was affected; there were to be no possibilities for private individuals to initiate action to protect a public interest. Development of implementative guidelines was not a matter for public participation but was limited to a small number of special interest groups. The channels open to affected parties were few and the possibilities to initiate action limited.

The main arguments used in the debate centered around the concepts of efficiency and effectiveness. Majority politicians used such arguments to support proposals to limit participation. As a result of environmental reforms, administrative effectiveness would increase considerably in coming years. The more effective the agencies became, the better able would they be to take care of both public and individual interests. In fact, strong agencies were seen as an adequate means to protect private interests, thus making increased public participation unnecessary. Minority arguments to the effect that public participation would be a valuable complement to administrative action and would lead to more effective implementation were not accepted. Neither did the majority explicitly discuss the problem of administrative accountability and popular control or the general value for democracy of public participation.

The United States choice was to extend the possibilities for public participation in the implementation process. All citizens would be able to participate, not only to take care of their individual interests, but also to protect public interests. The scope of participation embodied the development of implementative guidelines and standards, as well as the enforcement of such standards in individual pollution cases. The channels open to citizens involved administrative processes as well as proceedings in court. Any citizen would be able to initiate court proceedings, either to initiate a judicial review of implementative standards and guidelines or to secure the enforcement of such standards and guidelines in individual cases. Initiating enforcement to protect a public interest would not restrict the citizen's right to compensation and relief for individual damage suffered from pollution.

The arguments concerned not only efficiency and effectiveness but also popular control of administrative performance and the general value of participation in a democratic polity. Senator Muskie argued that limiting public participation could only be perilous to a society of government of and by the people. Citizen involvement would be a valuable complement to administrative action, and help to increase the efficiency and effectiveness of policy implementation. Citizen suits against the EPA administrator were also defended as a means of securing administrative efficiency. However, control arguments might have been more central in this case. Compelling bureaucratic agencies to carry out their duties is integral to a democratic polity. Therefore, citizens should have the right to control administrative performance and to intervene when agencies fail to perform their duties.

Arguments against increased participation concerned efficiency and control. For example, increased popular control was unnecessary; there was no reason to believe that administrative agencies would not do their utmost to implement public policy. Increased citizen involvement would decrease efficiency. Agencies would be locked up in numerous suits with no time to develop policies. Differing court opinions would make policies inconsistent and delay goal achievement.

The Swedish choice may be characterized as one of limiting public participation to secure a consistent and efficient administrative implementation of public policy. Performance was more important than participation; in fact, good performance would make participation almost unnecessary. The United States choice may be characterized as one of extending public participation to increase administrative efficiency and secure an adequate administrative performance. Participation was as important as performance; in fact, participation was necessary for performance.

CHAPTER 6

Explaining the Differences in Policy Choice

Physical and Environmental Considerations

Our study has revealed considerable differences in the content of clean air policies in Sweden and in the United States. The study has also shown that the arguments and motivations for and against different policy measures differ substantially between the two countries. In Sweden, policymakers view clean air as one among many competing and legitimate social objectives, believing that control efforts should take into account the benefits accrued from continuing air polluting activities. A major part of policy activity concerned the problem of controlling stationary source pollution. United States policymakers rather suddenly came to view clean air as an objective over and above other social objectives. In particular, the control of automobile-engendered pollution to protect public health was given top priority, and pursued seemingly without reference to socioeconomic consequences.

The explanation to be developed in this chapter could be formulated in this way: *The differences in policy content are a result first and foremost of the different political considerations of Swedish and American clean air policymakers. Such considerations influenced and governed the policymakers' physical-environmental, technological, and socioeconomic considerations.*

As a first effort to show the strength of this explanation, let us look at the character and importance of the policymakers' physical-environmental considerations—their perceptions of the air pollution situation as well as of the needs for and consequences of air pollution control measures. Let us assume that the differences in policy choice are all a result of considerations concerning the character, severity, and trends of air pollution in each country. In other words, policy content reflects a rational assessment of the pollution control situation and probable effects on that situation of particular policies.

In reporting its bill on September 17, 1970, the Senate Public

Works Committee viewed the bill as "the result of deep concern for the protection of the health of the American people." This concern had "increased since the publication of air quality criteria documents" that indicated that "the air pollution problem is more severe, more pervasive, and growing at a more rapid rate than was generally believed." Carbon monoxide levels, the committee reported, were in many places directly damaging to health, and a massive attack must be launched. The committee decided that public health was more important than technical and economic feasibility; "national air quality standards are authorized because the committee recognized that protection of health is a national priority." This sense of urgency is evident also in Senator Muskie's opening statement on the Senate floor: "It is a necessary bill, because the health of our people is at stake. . . . The costs of air pollution can be counted in death, disease, and debility. . . . Unless we recognize the crisis and generate a sense of urgency from that recognition, lead times may melt away without any chance at all for a rational solution to the air pollution problem."[1]

The threat to public health was used to motivate national ambient air quality standards to be reached within specified time limits. It was also used to promote the tough auto emission guidelines, which would lower harmful auto exhaust emissions from 1975–76 and later models by as much as 90 percent.

Swedish policy documents do not convey this sense of urgency. The 1969 cabinet proposal stated that the social and economic development has "far-reaching repercussions on environmental quality. . . . However, air pollution has not reached a precarious level, but concerted action is needed to prevent a dangerous situation." Such action would be taken to "look after environmental quality in the stiffening competition with other social objectives" and to "prevent damages of general importance." Of great importance to policy choice was also the perception that the Swedish pollution picture was unique; foreign standards and regulations could not be transferred and applied to Sweden.[2] As shown in chapter 3, these perceptions guided the adherence to the principles of technical practicability and economic feasibility, which, in turn, motivated the choice of individual source control as the major policy approach.

Does this indicate that physical-environmental considerations were indeed decisive in forming Swedish and American policymakers' choice? At first glance, one might be tempted to accept such a conclusion. On second thought, however, one might want to go further. Decisions are seldom based on considerations of one set of relevant factors alone; judgments based on one set of factors are linked to considerations of other sets of factors. Whether this is done deliberately or not, considerations of other relevant facts and arguments may lead decision makers to assess objectively similar conditions differently. Physical-environmental considerations must, therefore, be judged relative to the consideration of other factors, before a final judgment concerning their importance can be made.

As a first effort toward putting physical-environmental considerations in proper perspective, let us assess the tenability of the policymakers' physical-environmental arguments. Was America facing an air pollution crisis? And were Swedish air pollution problems as limited in scope and impact as the policymakers said?

The Extent of Pollution

In 1969, the average Swede was exposed to roughly two pounds of air pollutants a day (see table 4). Fifty-five percent originated from stationary sources, and 45 percent from mobile sources. In the same year, the average American was exposed to roughly seven pounds of air pollutants every day (see table 5). Sixty percent of the pollutants emanated from mobile sources, and 40 percent from stationary sources.

However, the weight of air pollutants is not the only relevant measure of air pollution. One must particularly pay attention to the effects of each pollutant. A ton of sulfur dioxide means more than a ton of particulates or carbon monoxide, because sulfur dioxide is more damaging to health. Because of Sweden's frosty climate, and the heavy dependence on oil for energy production and heating, the single most important pollutant in 1969 was sulfur dioxide. In wintertime, emission levels in the Stockholm and Gothenburg areas were regularly above those recommended to safeguard public health. The problem was—and is—compounded by the prevailing winds over Europe, which bring large amounts of "foreign" sulfur to be precipitated over Swedish territory.[3]

Sulfur dioxide was a serious problem also in the United States; in both countries, the average citizen load of that pollutant was about 250 pounds a year. But it is also true that mobile source pollutants, such as hydrocarbons and nitrogen oxides, constituted a much more serious problem in the United States. The average American was exposed to five times more of these pollutants than the average Swede. Furthermore, these pollutants under certain weather conditions react with each other to produce photochemical smog, which has very serious health effects. Finally, it is worth mentioning that the citizen load of airborne lead from auto exhausts may have been four times higher in the United States than in Sweden in 1969.[4]

Admittedly, these data do not seem to contradict the conclusion that physical-environmental considerations were decisive in forming the policy response in the two countries. The perceptions of Swedish policymakers that air pollution was not yet a serious problem in need of

TABLE 4. Estimated Amounts of Air Pollutants by Type and Source in Sweden in 1969 (in millions of tons)

Source	CO	Particulates	SO_x	HC	NO_x	Total	Percentage distribution
Transportation	1.07	0.02	0.006	0.162	0.044	1.302	44.9
Fuel combustion in stationary sources	0.05	0.009	0.415	–	0.1	0.574	19.8
Industrial processes	0.06	0.37	0.476	0.034	0.055	0.995	34.3
Solid waste disposal	0.02	0.001	0.003	0.004	0.001	0.029	1.0
Total	1.2	0.4	0.9	0.2	0.2	2.9	
Percentage distribution	41.4	13.8	31.0	6.9	6.9		100

Sources: Sweden's national report to the United Nations on the human environment (1971), p. 35; SOU 1974:101, pp. 70 f.; SOU 1975:98, p. 22.

TABLE 5. Estimated Amounts of Air Pollutants by Type and Source in the United States in 1969 (in millions of tons)

Source	CO	Particulates	SO_x	HC	NO_x	Total	Percentage distribution
Transportation	111.5	0.8	1.1	19.8	11.2	144.4	60.1
Fuel combustion in stationary sources	1.8	7.2	24.4	0.9	10.0	44.3	18.4
Industrial processes	12.0	14.4	7.5	5.5	0.2	39.6	16.5
Solid waste disposal	7.9	1.4	0.2	2.0	0.4	11.9	5.0
Total	133.2	23.8	33.2	28.2	21.8	240.2	
Percentage distribution	55.5	9.9	13.8	11.7	9.1		100

Source: U.S. Council on Environmental Quality, *Environmental Quality: The Second Annual Report of the Council on Environmental Quality*, (Washington, D.C.: Govt. Printing Office, 1971), p. 212. NOTE: the category of "misc. sources" is omitted here.

drastic action seem corroborated by available data. Correspondingly, the United States data seem to justify what Senator Muskie called "a stern response." Furthermore, the relatively more serious stationary source pollution in Sweden and the relatively more serious mobile source pollution in the United States seem to be accurately reflected in the policymakers' perceptions and responses in the two countries.

But this seems to indicate the outer limits of the physical-environmental line of explanation. Other characteristics in the pollution picture, such as future air pollution trends, seem to suggest a more modest role of physical-environmental perceptions and preferences in policymakers' considerations.

In Sweden, the number of motor vehicles doubled during the 1960s, and there were signs of a further increase in coming years. Already in 1965 the energy consumption per capita was higher than in any other European country. Moreover, 85 percent of all energy consumption was based on fossil fuels, and this figure was increasing at a rate that indicated a doubling in fifteen years. Clearly, the air pollution situation would soon become much worse.[5]

The American prognosis looked quite similar. The number of motor vehicles on the road was expected to double by 1980. Already burning more energy than Great Britain, Japan, the USSR, and West Germany together, it was projected that the United States would consume more energy between 1970 and 2000 than in its entire earlier history. Given the available sources of energy, these trends indicated the need for a rapid switch to high-sulfur, and thus highly polluting, coal as a major means of meeting the immense demand for energy.[6]

In the face of these predictions, why did United States policymakers make a drastic overhaul of national priorities while the Swedes did not? Had physical-environmental considerations of this kind been decisive, the responses should have been more alike. But only the United States policymakers perceived a future pollution scare as justifying drastic measures today; the Swedes seemed confident that incremental and feasible policy steps would be sufficient to meet the air pollution problems of their nation.

These choices of policy content become all the more curious if one looks at the short-term trends of air pollution. Figures revealed in 1967 indicated that Swedish air pollution would increase by 50 percent between 1965 and 1969. If anything, this would have justified a more radical approach, had physical-environmental perceptions and preferences been uppermost in the policymakers' minds. Evidently, this was not the case. As for the United States, almost all measures used indicate an actual improvement of air quality between 1969 and 1970. For two of the five major pollutants, 1970 emissions were less than those of 1969; emissions were larger for only one of the major pollutants. The total amount of air pollution had changed only slightly since 1967, when the earlier feasibility approach was chosen. Muskie's switch from an incrementalistic feasibility approach in his first bill of March 4, 1970, to his radical stance in August and September culminating, finally, in his remarks on December 18, 1970, that "there was little doubt in the Senate, in September, that the country was facing an air pollution crisis," could thus not be justified solely by physical-environmental considerations. Evidently, Muskie and other United States policymakers must have been motivated by other considerations when proposing such drastic policy measures as those contained in the 1970 Clean Air Act Amendments.

Two conclusions can be drawn from this analysis: (1) physical-environmental considerations played an important role in policy choice, but (2) this importance was limited to certain aspects of the choice. As one could have expected beforehand, the main thrust of the selected alternatives accurately reflected the relative importance of stationary and mobile source pollution in each of the countries. Furthermore, the relatively more serious overall pollution picture in the United States was reflected in the more drastic and far-reaching measures chosen in that country. Correspondingly, the not yet alarming air pollution situation seems to have been a major motive for Swedish policymakers to recommend incremental and practicable policies rather than an all-out attack to protect public health once and for all from the threat of air pollution.

But neither the drastic United States reordering of national priorities nor the Swedish policy marginalism could possibly have been motivated solely on physical-environmental grounds. Projections of future air pollution trends and consequences did not differ that much, and short-term trends would have motivated policy choices contrary to the ones actually made, had environmental concerns been uppermost in the policymakers' minds. The same is true for the distribution of policy powers. A consistent environmentalist approach would have supported continuing along the line of establishing air-shed-based air quality control regions. Instead, the United States approach now became much more geared to establishing standards and distributing powers in accordance with existing political, jurisdictional, and administrative boundaries.

It thus seems safe to conclude that policymakers in both countries perceived other objectives, consequences, and arguments as equally important or more important reasons for making their choice of one particular policy alternative. The question becomes which ones were more important, and what were the linkages among the physical-environmental and other considerations relevant to the choice of clean air policy.

Technological Considerations

As we have seen in earlier chapters, politicians in the two countries differed widely in their assessment and choice of approaches to pollution control technology. American policymakers seem to have been driven by their concern for the protection of public health; very stringent air quality standards and emission standards were established to speed up and force development of not yet available, but necessary, control technologies. Swedish policymakers seem to have been motivated more by their concern with orderly and planned socioeconomic development, adjusting pollution control requirements to present levels of technological development and changing requirements as new technologies became practicable.

This gives us an indication of the very special character of technological considerations. Only in a very general sense is technological development perceived as a goal in itself. In the majority of decisions, it is considered a means for achieving other goals and realizing other preferences. It follows from this that judgments of the availability, costs, and consequences of a particular technology differ depending on what circumstances and objectives are uppermost in the policymakers' minds at the time of choice. Considerations of technology per se do not explain policy choice; however, they are linked to other considerations and may thus provide a clue to what was decisive in forming the policymakers' decision.

This line of reasoning is vividly illustrated by the different approaches to technology taken by the two houses of Congress. In June 1970, the House of Representatives was still proposing a policy of adjusting pollution control standards and measures as new control technologies developed. As Committee Chairman Staggers put it: "Effective technologies to reduce or eliminate particular pollutants must be developed. Many people think these have already been developed and can be put into effect, but this is not true. That is the reason why almost half the money in this bill is for research and development." The House committee hoped that the two great industries—automobile manufacturers and automotive fuel producers—would join hands to develop such effective technologies. In September 1970, the Senate statutorily forced the great industry of automobile manufacturers to install in all cars produced five years later a control technology that at the time of choice was projected to be available in two prototypes by 1975! The reason for this drastic shift in approach thus was not a proved and practicable breakthrough in technology. Senator Muskie said that "the deadline is based not, I repeat, on economic and technological feasibility, but on considerations of public health. . . . We are saying that Congress, in the interest of public health, should say to the country and to the industry that this is what the health of this nation requires."[7] As was shown above, the short-term trends of air pollution did not indicate any drastically increasing threat to public health; air quality had not "changed enough to produce a crisis in September that was not apparent in March," when Muskie still seemed content with an adjustment approach.[8] Evidently, Muskie and others at that time perceived other circumstances, consequences, and objectives as important enough to warrant a dramatic shift in policy content.

In Sweden, the attitudes toward technology, and thus the role of technological considerations, never changed. Throughout the long and scattered process of choice, practicability and economic feasibility remained the main criteria for assessing the necessity and acceptability of different pollution control alternatives. The availability of control technologies and the economic consequences of adopting new technologies were considered important constraints on policy choice. Long-term commitments to programs for regulating the sulfur content in fuel oils were rejected on the grounds that measures should be taken only at the pace permitted by practical and economic conditions. Mobile source control measures could be taken only when "a technological base for desired action has been developed," said the minister of transport, Olof Palme, in 1968. Great care was taken to assess the economic consequences of stationary source control technologies required by different emission standards, and a system of subsidies was developed to help those firms which would otherwise have had to shut down as a result of pollution control requirements; the joint NEPB-industry expert panels were instrumental in "concentrating the discussions, or negotiations, on economic feasibility, i.e., the heart of the matter."[9] In contrast, the United States Senate argued that "existing sources of pollutants either should meet the standard of the law or be closed down."[10]

In no way do these differing approaches to technology reflect different levels of knowledge concerning the latest developments in the field of pollution control technology, or the costs and benefits of different alternatives. Royal Commission reports and specialist panel reports, as well as Congressional hearings, abound with material concerning the development and availability of control technologies throughout the industrialized world.

What emerges here are two distinct patterns of considerations and policy choice. In Sweden, policymakers perceived the environmental situation as one that should be carefully watched. However, the Swedish view was that the situation did not warrant drastic action, and environmental quality should not take priority over other social objectives. Pollution control measures should be adopted gradually, as new technologies were proved practicable and economically feasible. In no way should measures be taken which could jeopardize the achievement of such objectives as full employment, economic and social welfare, and international competitiveness. In the United States, policymakers considered air pollution an immediate threat to public health. This seemed to motivate strong action, regardless of the socioeconomic consequences, and was accompanied by a strong faith in a technological "fix"; technologies would be forthcoming if only the policymakers told industry to develop them.

We have shown that the actual air pollution picture in the two countries in itself did not motivate these patterns, nor did knowledge of pollution control techniques differ enough to warrant such different policy choices. Instead, we are left with the suspicion that consideration of other factors relevant to the choice governed and formed the choice of policy. Technological considerations only reflect the importance of these other perceptions and preferences; they do not explain why Swedish policymakers kept and American policymakers changed their views on air pollution control. Evidently, further looks are needed.

Socioeconomic Considerations

It thus becomes important to find out why Swedish policymakers stuck to long-held objectives of socioeconomic welfare and stable growth and never gave priority to environmental quality or public health, while American politicians switched to judging public health and environmental quality as the most important social objectives. Since we assume that our policymakers are rational actors, capable of assessing the substance matter confronting them, we may further assume that their assessments correctly reflected the actual socioeconomic conditions and consequences relevant to clean air policy alternatives. If we further assume that American and Swedish policymakers gave equal weight to socioeconomic considerations, we may hypothesize that American policymakers found that socioeconomic costs of radical policy measures were outweighed by health and environmental benefits, while the Swedes found the costs larger than the benefits. But if they seem to have judged similar socioeconomic conditions and consequences differently, then we will have to look further for a more valid explanation.

To Muskie and the Senate, it had been clear since early June that the "law could result in drastic changes in the pattern of the life we live in the urban areas of America. . . . It would require new discipline of our desire for luxury and convenience." But the costs of some plants closing down, of restrictions on private car transportation, and of regulating the use of property would indeed be worthwhile; it would save billions of

Explaining the Differences in Policy Choice   111

dollars of property losses caused by air pollution, and it would save billions of dollars in "the annual costs of health care in America."[11]

How different the view of Herman Kling, the Swedish minister of justice: "The competitiveness of Swedish industry must not be weakened as a result of too rigorous pollution control requirements. . . . In view of the socioeconomic consequences of factory close-downs, a plant should be forced to close down only if the costs of pollution so outrate the benefits of continuing production that it must be considered indefensible to allow the plant to continue."[12]

As we have seen earlier, these different views prevailed at the time of choice in each country. Both views, perhaps especially the Swedish one, point to the existence of strong preferences guiding the policymakers' assessment of the acceptability of the socioeconomic consequences of pollution control measures. But before we conclude that this was the case, and turn to the problem of how these preferences emerged or prevailed, let us see whether the different assessments of socioeconomic consequences were related to actual socioeconomic differences and perceptions of those differences.

In particular, we want to find the hard facts that the policymakers in the two countries might have had at their disposal at the time of choice. What were the costs of air pollution at the time? What industries would be affected by strong anti-air-pollution measures, and what groups? What would be the ramifications for energy production and supply and for transportation of clean air policy measures? What would such control measures cost?

Estimates of the damage costs of air pollution vary considerably, depending on what is included. Figures released to the public in 1968 indicated that air pollution damage to buildings, equipment, and vehicles cost the average Swede $32 every year, and the figure was climbing. If to that were added the costs of decreased forest growth caused by sulfur in air and precipitation, the per capita cost was probably up to or above $36.[13] A report released by the United States National Air Pollution Control Administration in 1968 estimated the damage costs from air pollution to be about $41 per capita.[14] The difference does not seem large enough to warrant the different judgments and choices made by policymakers in the two countries.

But did the manufacturing branches contributing most to air pollution perhaps have a more limited socioeconomic importance in the United States than in Sweden, thus making it easier for American policymakers to play down the socioeconomic consequences of strong pollution control measures? As can be seen from table 6, this was hardly the case. The five main air polluting manufacturing branches in both countries represented about one-seventh of manufacturing plants,

TABLE 6. Employment, Number of Plants, and Value of Shipments for Main Air Polluting Manufacturers

Manufacturing Branch	Sweden			United States		
	Number of Plants	All Employees (1,000)	Value of Shipments (billions of Skr)	Number of Plants	All Employees (1,000)	Value of Shipments (billions of $)
Pulp and paper	305	56.3	7.0	5,903	638	20.7
Chemical, petroleum, and allied products	666	47.4	6.0	13,691	991	62.5
Stone, clay, and glass	1,180	46.2	3.2	15,602	591	14.6
Primary metal industries	216	65.2	6.9	6,821	1,279	45.5
Motor vehicles and equipment	149	28.9	3.5	2,652	743	40.4
Total for these branches	2,516	244.0	26.6	44,669	4,242	183.7
Percentage of all manufacturing	15.2	25.8	32.0	14.4	22.0	32.9

Sources: Statistisk Årsbok för Sverige 1969, SOS: Industri 1967; Statistical Abstract of the United States 1970. NOTE: auto industries are not in themselves polluting, but will be affected by regulations to control auto pollution.

around one-fifth of all employees in manufacturing, and one-third of the value of shipments from all manufacturing. Air pollution control measures against industrial polluters could thus have potentially the same socioeconomic consequences in both countries.

Still, the United States policymakers swapped earlier policies of feasibility and incrementalism for an all-out effort to wipe out pollution in a few years. Polluting sources that could not meet the standards set to protect public health must be closed down. As we have seen above, industrial pollution had changed only marginally from the earlier policy choice in 1967–70 and so had industrial structure. On substantial socioeconomic grounds, there seems to be no evident motive for the drastic change in preferences. Given the fact that the socioeconomic importance of polluting industries was roughly the same in the two countries, one is led to wonder why the Swedes did not take tougher measures, since industrial air pollution proper was relatively more important in Sweden.

The benefits of cleaner air may be especially appreciated by people in urbanized areas. However, air pollution control measures also inflict costs; auto emission standards and emission standards for stationary sources may result in higher prices for travel and energy. Did the situation with regard to transportation, energy production, and urbanization differ enough in the two countries to account for the differences in policy choice?

By 1965, two-thirds of all Swedes lived in urban areas with more than 2,000 inhabitants. Almost 90 percent of all passenger travel was by private cars, and 40 percent of all transportation of goods was by motor vehicles.[15] In the same year, two-thirds of all Americans lived in areas with more than 2,500 inhabitants. About 90 percent of all passenger travel was by private cars, and 22 percent of all transportation of goods was by motor vehicles.[16]

With regard to energy production, Sweden seems to have been somewhat more dependent on fossil fuels; with most of the rivers exploited, the limits of hydroelectric power were in sight, and foretold an even stronger dependency on imported fossil fuels (see table 7). In the United States there was a high dependency on fossil fuels for electric energy production. The dependency on high-sulfur coal was projected to increase.

The figures seem to indicate that Swedish and American policymakers were facing quite similar socioeconomic influences on policy choice. Now, I am the first to admit that air pollution was worse in the United States than in Sweden. After all, New York is dirtier than Stockholm and Los Angeles's problems do not exist in windy Gothenburg. This could motivate a strong United States line against polluters, with a somewhat less-pronounced attention paid to

TABLE 7. **Mineral Energy and Electricity Production in Sweden and the United States in 1965 (in percentage of total)**

Energy source	Sweden	United States
Coal, oil	72.0	60.6
Natural gas	–	35.2
Electricity	18.0	4.2
Other	10.0	–
Total	100.0	100.0
Main source of electric energy		
Hydro	87.5	18.4
Coal, oil	12.5	60.6
Gas	–	21.0

Sources: SOU 1970:13; *Statistical Abstract of the United States 1970.*

socioeconomic consequences. On the other hand, we have seen that neither the amount of pollution nor the socioeconomic importance of polluting industry, transportation, or energy production had changed more than marginally since the earlier feasibility approach was adopted in 1967. Why was a majority of United States legislators ready to challenge such enormously important socioeconomic forces as the car-manufacturing and oil-producing industries by choosing to protect public health at any cost in 1970? Furthermore, if there is less pollution, there should also be less socioeconomic impact from pollution control measures. Why then were so many Swedish policymakers so anxious to stick to the criterion of economic feasibility in 1969?

Part of the answer to these questions may be found in the policymakers' assessment of some socioeconomic differences that, of course, exist between the two countries. Representing a country of 8 million people, with about two-fifths of industrial production sold on a very competitive international market, Swedish policymakers are forced to take a hard look at the consequences of policy measures that place burdens on industry. As we have seen, the main argument for proceeding with auto exhaust and fossil fuel controls only as the capacity to bear additional burdens existed was motivated by the international competitiveness of Swedish industry. The majority view was succinctly summarized by the minister of agriculture in his proposal for subsidizing pollution control measures in existing industrial plants: "In principle, I favor the view that pollution control costs should be part of the costs of production. But I am aware that control measures could be a heavy burden for certain industries. As long as we do not have international standards for environmental protection, measures taken by export industries in one country could jeopardize their international competitiveness which in turn could have a negative effect on employment."[17]

The United States is the world's leading industrial nation. Its oil-producing and car-manufacturing industries are giants in international comparisons. At the same time, only about 7 percent of the American economy is dependent on foreign trade. Because of the size of the American economy, this 7 percent still represents a very important part of the total international trade. In other words, American policymakers may have a different set of socioeconomic conditions to take into consideration when making their policy choices.

This can be seen in the reasons given by Nixon and shared by other policymakers for establishing national air pollution control standards. The competitiveness problem was internal rather than international; the important thing was to eliminate the possibility that some states would set laxer standards and thus establish pollution havens. But Muskie foresaw no socioeconomic problems from setting tough national standards. American economy and American industry had been "presented with challenges in the past that seemed impossible to meet, but made them possible." Given their successful record of the past, such as the war economy and the moon program, Muskie found it "difficult to believe that, whatever their present doubts, they cannot meet the challenge of this bill."[18]

One is led to wonder whether indeed man's first step on the moon in 1969 made possible in 1970 what had been impossible in 1967. As we have seen, Senator Griffin was not easily convinced, arguing that manufacturing, sale, and servicing of motor vehicles was still a vital industry in the United States economy. And that industry was not without problems; Griffin warned that the bill would saddle the automobile industry with unreasonable demands and expenses at a time when rapidly rising costs were already putting it under severe handicap in competing with foreign producers.

Usually, such arguments would have been decisive in forming the policy choice on Capitol Hill. Why not in September 1970? At the end of his speech, Senator Griffin put his finger on a possible explanation: "Frankly, I think one of the problems with this legislation right now is that . . . too many of the decisions with regard to this bill are being made on a political basis."[19]

Senator Griffin was right; there must have been other considerations that formed policy choice, for the actual pollution pictures, the developments in pollution control technology, and the socioeconomic consequences did not seem to warrant the wide differences in priorities and policy choice. Still, the policymakers in the United States perceived the situation as having changed dramatically enough to necessitate drastic action with less than usual regard to socioeconomic consequences and technological capability, while Swedish policymakers perceived the situation as necessitating nothing but limited and "reasonable" action within the nation's economic and technological capacity. To find out what considerations were really decisive in changing the American policymakers' pattern of choice and what considerations really made their Swedish counterparts stick to one approach, another look is needed.

Political Considerations

The time is obviously ripe to let the hare and the tortoise come out into broad daylight—especially since we may now suspect that it is not the outdoor track but the runners and their way of running that makes the most conspicuous difference. If our argument so far is correct, it follows that the most satisfactory explanation is also the one that should come most easily to the political scientist's mind: the differences in policy choice are due to the policymakers' assessments and evaluations of the political costs and benefits of different policy alternatives.

The political assessment concerns both substance and strategy. The policymaker must judge the acceptability of a policy alternative in terms of its consistency with existing constitutional and institutional arrangements; political costs and benefits for the same policy alternative differ between federal and unitary systems. Policymakers must also evaluate alternatives in terms of their consistency with generally held cultural norms and beliefs. They must judge the acceptability of policy alternatives in terms of central long-term ideological objectives and policy commitments. From the strategic point of view, policymakers must judge policy alternatives in terms of their impact on such objectives as personal popularity, coalition building, extended political power, and improved relations with particular interests.

The starting point of the analysis, then, is that the political costs and benefits for similar alternatives are not the same in the two countries. Constitutional and institutional arrangements, political objectives anchored in prevailing political cultures and ideological beliefs, and current public opinion and party and interest group relations provide policymakers with different obstacles and opportunities of choice. What we assume is that the choices are consistent with a rational assessment of these different political costs and benefits.

My main proposition is that in times of a large upsurge in public opinion, the constitutionally built-in competitiveness of the American political system coupled with the peculiarities of the electoral and party systems will provide policymakers with tremendous incentives for selecting drastic and escalatory policies. Because they are so visible in the eyes of their constituencies, but with no responsibility for actual policy implementation, the United States Congressmen will see much merit in selecting radical alternatives. At the same time that constitutional arrangements as well as prevailing patterns in political culture prevent them from running too fast in such areas as centralization of policy powers, these same factors will encourage them to increase their speed in areas of public participation.[20] Less visible to his constituency because he primarily represents his party, which in the future might be part of a governing coalition and thus responsible for policy implementation, the Swedish Riksdagsman will see much merit in incremental policymaking. At the same time, prevailing organizational and cultural patterns will induce him to centralize policy powers and to keep participation at a level that does not upset the gradual but steady progress toward chosen policy goals.

Evidently, the status and trends of public opinion are most essential to this argument. We have always been taught that the United States Congress is the home of bargaining and compromise—"so much so that even the best idea, with broad appeal, cannot survive intact."[21] But we are also told that "the normal gradualism is not only consistent with but conducive to periodic swings into extreme commitments of American public policy. In fact, many of the incremental adjustments in American policy are corrections that have been required by such abrupt, unpremeditated commitments."[22]

How broad and conducive was the environmental appeal in 1969-70? Indeed, there was a dramatic upsurge in American environmental opinion around 1970. A June 1970 poll indicated that nearly 70 percent thought air pollution was a serious problem in their own area and that 50 percent thought it had grown worse in latter years. Polls in 1969-70 showed a majority of 54-56 percent in favor of increased federal spending on pollution control, even if that would mean an individual tax of up to $15. An October 1970 Gallup Poll indicated that 58 percent perceived a candidate's stand on pollution control as "extremely important" to the electorate's mind in the upcoming election. And in November Gallup reported that up to 79 percent of the electorate favored having all new cars equipped with an antipollution device that would add about $100 to the price of the car.[23] In Washington, pressure was growing on politicians to do something about pollution. The number of environmentally oriented organizations with Washington offices grew from forty-five to sixty-seven in two years. During 1969-70, the number of registered "conservation" lobbyists grew from two to thirteen, and some of the new lobbyist groups used quite militant tactics in getting their message across to the politicians. The Earth Day in April 1970 was perhaps the most impressive environmental demonstration ever.[24]

This upsurge in environmental opinion in an election year clearly affected the process and content of policy choice. As a nationally elected politician, the American president is expected to be especially sensitive to trends in public opinion and apprehensive of the activities of future opponents. The polls indicated that middle-class, middle-aged suburbanites and the young generation were the most interested in environmental matters. For a political animal like Richard Nixon, this was an opportunity of sorts. Not only did it provide him with a timely theme for his 1970 State of the Union Address and a means of diverting the people's interest from the Vietnam war, it also gave him a chance to deal a blow to one of his most important opponents for 1972. As the Democratic candidate for vice president in 1968, Edmund Muskie was considered a front-runner for the 1972 Democratic nomination. Much of his political fame was earned from his leadership in environmental affairs.

Hence Richard Nixon's sudden—and as it turned out, rather short-lived—interest in pollution control. His proposals of February 1970 were quite radical compared to the existing "Muskie" policy, and clearly gave him the initiative. Muskie's leadership was further questioned by the immediate bipartisan House support for the Nixon proposals. In early May, things would get even worse for Muskie. Then a Ralph Nader report on air pollution strongly criticized Muskie's role in the past and just stopped short of accusing him of selling out to industrial polluters. This strong criticism of a presidential hopeful was excellent stuff for the media, which gave the report wide coverage.[25]

If Muskie was to retain his leadership and improve his image as "Mr. Environment," his strategic position in the spring of 1970 seemed to leave no options open but to enter the game of policy escalation. Proposals more radical than those of the president and the House would no doubt serve him well in his own reelection campaign. Developments between May and September 1970 confirm the paramountcy of strategic considerations. In June—the day before the House passed its bill!—Muskie held a press conference indicating support for very tough auto emission regulations. Later in that month he said provisions far beyond anything proposed so far were being considered in his subcommittee. As we have seen, this was not empty rhetoric. The Senate put public health above economic feasibility as the main policy objective, and put industry under pressure to come up with solutions, seemingly disregarding costs. The House went along with most of the Muskie bill, and Nixon finally signed it. Muskie had recaptured his position as a leading environmental policymaker.[26]

The arguments used in defending this radical approach clearly reflect the importance of strategic considerations. Time and again policymakers provided variations on the same theme: "Public concern about air pollution is so great that there is a groundswell of support for strong legislation." And Muskie frankly admitted that "in the climate of environmental concern which we faced in the country, it was important that Congress give to the country the best bill it was possible for Congress to devise." Senator Griffin was right; when "everybody is for clean air and against pollution [it is indeed] difficult politically to vote for any amendment that would be characterized by the press as weakening the bill."[27]

Hence the swing from economic feasibility to health protection as the central policy objective, and hence the emphasis on national standards and expansion of federal powers, as well as the seeming neglect for the technological and economic objections from the traditionally very strong industrial interests, lobbying hard on Capitol Hill.

But why were Senators and Representatives so sensitive to public opinion? As will be shown below, Swedish policymakers faced the same environmental concern, but did not escalate policy.

To understand why policy escalation under certain circumstances can become such an attractive strategy for American policymakers, one must take a look at the political context, i.e., the constitutional, institutional, and party systems in the United States. The Senator and the Representative do carry party labels, but party identification is not as important as in a parliamentary system. The electoral system makes the policymaker a representative of his constituency rather than of his party. The candidacy and the campaign are based on personality rather than on party affiliation. Consequently, the candidate's or the elected policymaker's personal views and standpoints on issues come under direct and continuous scrutiny by constituents. Candidates and elected officeholders are thus provided with very strong strategic incentives for following public opinion and public concern.

The balance-of-power system of government provides the leading policymakers with great opportunities to compete for and grab the political initiative. They can arrange hearings to get crucial information independently of the executive and the bureaucracy and—not of least importance—promote their own policy ideas. In times of rapidly growing public concern with a particular issue, the system thus provides the Congressman with tremendously strong incentives for policy escalation rather than incrementalism. This is even more so because Congress will never be held responsible for the implementation of the legislation it passes. Should it turn out that radical legislation has not had the intended impact, Congress can always blame the executive and the bureaucracy for bad performance. This was exactly what Muskie did to defend and at the same time dissociate himself from the evidently very unsuccessful 1967 air quality legislation, and the House Committee began its report with a strong attack on the federal air pollution control agency.[28]

But policy escalators in America must also take into account the policy alternative's costs and benefits in terms of political substance. Political beliefs in America are predominantly against dramatic increases in government responsibilities; " . . . the machinery of government is not an accepted piece of institutional apparatus to be made use of as and when required; it is a sort of emergency appliance, to be wheeled out only in the most extreme circumstances and put back in its place, if at all possible, as soon as the emergency is over."[29] Such beliefs go hand in hand with constitutional and legal tradition. The Constitution carefully seeks to strike a balance between the needs for political government and individual liberty, and provides the citizen with judicial means to control and check governmental performance. This strong position of the judicial branch both affects and expresses the strong cultural belief in due process and the rule of law. The rights of states and local communities place another constraint on policy escalation. Thus, when the dominant idea is that government is on a leash held by the people rather than the other way around, it follows that every proposal for policy escalation—however necessary it may be perceived by politicians and public opinion to "do something" about a particular problem—must be judged in terms of its consistency with constitutional and institutional principles and features and with dominant political beliefs.

In picking increased public participation as the first in his series of escalatory moves, Edmund Muskie thus accomplished several things. Strategically, he provided a carrot for environmental groups, who wanted to clean up America. Muskie also seized the political initiative by proposing a policy of citizen involvement in agency actions that the president, as chief executive, would hardly like, but one which he could not actively oppose, given prevailing political beliefs. In terms of substance, Muskie made his choice consistent with the mainstream of American political culture by alluding to classical arguments: " . . . in a system of government of and by the people, we foreclose public participation at our peril."

Confronted with a choice between performance and participation, the American policymaker will thus be inclined to prefer participation. The institution of judicial review and the strong belief in due process and the rule of law all work toward inducing the policymaker to feature policy proposals in line with this preference.

The strength of constitutional, institutional, and cultural considerations are seen also in the choice of power distribution. Increased federal powers and national standards were motivated by the necessity to protect public health, a traditional federal concern. Proposals to allocate all standard-setting power to the EPA, as well as to give the EPA almost unlimited powers to substitute federal for state implementation plans and actions, were rejected not so much for reasons of impracticability (how could a central agency possibly control the hundreds of thousands of polluting sources around the nations?) as for reasons of precedence and the preservation of state and local powers. In voting for judicial instead of congressional review of administrative decisions to suspend auto emission deadlines, the Congress showed its preference for "the principle of due process which is embedded not only in our Constitution but throughout our legal system."[30]

We may thus conclude that political considerations played a very important role in forming American policymakers' choice. Because of the features of the electoral, party, and legislative systems, they had strong incentives to respond to the dramatically increasing public concern with environmental quality and pollution control. In Charles O. Jones's words: " . . . in 1970 a majority seemingly awaited unspecified strong action. Thus, instead of a majority having to be established for a policy, a policy had to be constructed for a majority. Much of that occurred within Congress as proposals escalated toward various actors' perceptions of what was necessary to meet public demands."[31] To this we must add that the response also reflects the policymakers' judgment of political substance. Policy escalation may have shown a neglect for technological or socioeconomic consequences, but certainly not for political ones. Policymakers did not change the distribution of powers or the possibilities for public participation beyond what they considered to be in line with the prevailing constitutional and institutional structure.

In Sweden, public concern with environmental quality and pollution control increased during the late 1960s. Polls released in spring 1969 indicated that 84 percent wanted to put charges on polluting industries. Four-fifths of the urban population, and 69 percent of all Swedes, were in favor of higher local taxes to fight pollution. As much as three-fifths favored prohibiting burning high-sulfur oil for heating and energy production. Perhaps the most remarkable figure was the 54 percent favoring a lower growth rate if that would save the environment. Fifty-seven percent wanted more attention to environmental problems in the mass media, despite the fact that the number of "environmental" editorials showed a sixfold increase after 1963.[32]

Clearly this growth of public concern provided incentives for policy escalation and a shift in priorities. To understand why incrementalism prevailed in clean air policymaking, one must take a closer look at the incentives provided by the institutional context of Swedish policymaking.

Sweden is a parliamentary democracy. Supported by a majority in the Riksdag, the party that forms the cabinet can count on getting its policy proposals ratified by the legislative body. The Riksdag provides limited resources for information-gathering independently of the cabinet and the bureaucracy. Hearings play a limited role, and the main activity is geared to cabinet proposals rather than private members' bills. The electoral and party systems provide a set of incentives quite different from those in the United States. The Riksdag member is only one of several representatives from each constituency. Candidates are nominated by the party organizations in the constituencies and present the party views in the campaign. Riksdag members are thus elected because of party affiliation rather than personality. They have more to gain from going along with the party leadership than from trying to establish themselves as champions for a particular group or issue. In a sense, they are therefore more insulated from public opinion than their American counterparts. It is the party leadership that must be aware of shifts in public opinion. But because of the parliamentary system, the party leadership must also be aware of the possibility that its party may form a cabinet in the future. It would then become responsible for the implementation of selected or recommended policy alternatives. This certainly provides an incentive for preferring the practicable to the desirable.[33]

What prevents this strong cabinet from exploiting issues and escalating policies? First of all, there is the ideological profile of the governing party. To the Social Democrats, ideologically committed to full employment and stable socioeconomic growth and welfare, environmental quality could never become *the* central issue. This long-term commitment influenced the terms of reference laid down for the investigatory Royal Commissions, which, in essence, meant narrowing down the scope of policy search and choice. Most of the time Royal Commissions consist of administrators and representatives of regulated interests. The latter also sit on the boards of administrative agencies responsible for implementing policy. Under such circumstances, it is not surprising that efforts to find out what is feasible and acceptable to affected parties play such an important role in Swedish policymaking. When policy proposals land in the Riksdag, they represent a compromise, backed up by such formidable forces as the cabinet, the bureaucracy, and the regulated interests. It seems safe to conclude that this context provides strong incentives for incrementalism and consensus around feasible alternatives.[34]

Seen in this light, the prevalence of feasibility and reasonableness as main policy criteria is easier to understand. The directives given to the Royal Commission in 1963 defined the problem of pollution as one of regulating nuisances from the use of property, that is, of regulating individual polluting sources. By putting the problem in terms of nuisance, the cabinet in effect narrowed down the area of policy choice. By the time public opinion began to demand environmental quality and the protection of public health against the damaging effects of pollution, the Royal Commission's report was generally accepted by the government and the regulated interests. Furthermore, discussions and negotiations between the NEPB and regulated industries had proceeded very far toward agreement on the implementation of the coming legislation. With so many powerful actors already committed to a particular policy alternative, there was little incentive for the private Riksdag member to try to break the policy open by proposing a shift in policy priorities. It is perhaps telling that only the Liberals, who are less tied to any particular interest group or class than any other Swedish party, presented a policy view resembling the one discussed in the United States.

The process of designing implementation guidelines clearly spells out the links between substance, preferences, and strategy. Responsible actors carefully explored the common ground of feasibility together with the industries that would be regulated. Care was taken not to suggest anything that would impose unreasonable burdens on Swedish economy and Swedish industry or imply policy programs beyond technological and administrative capability. Public health and air quality were mentioned more in passing, and then only as long-term goals. What was important now was to design a system of best practicable means within the capacity of regulators and regulated alike. Most policy actors seemed to agree that a small country like Sweden could not afford to gamble with industrial and economic development by establishing stringent air pollution control standards. By taking a small step now, it might be easier to ask for further steps at some later point. Indeed, the meeting at the Volvo foundries in Skövde presents the whole policy-making context and all the considerations of the policymakers in a nutshell.

Furthermore, the institutional context implies heavy political costs of efforts to spread policy powers or to increase individual public participation in policy implementation. Sweden has never been a federal system or a nightwatcher's state. The predemocratic patriarchal state was soon followed by the modern service democracy. The success of the state in providing welfare and other services helped foster a popular emphasis on what the policymakers do rather than how they do it, an emphasis that by the end of the 1960s had yet to be profoundly challenged. Sweden has its ombudsman, but the position of the judiciary is much weaker than in the United States. Judicial review of administrative performance is unknown to the Swedish system of government, and placing policy responsibilities in courts rather than in administrative agencies is, therefore, bound to meet with resistance. Furthermore, the idea of direct individual participation in the policy process outside a client status seems somewhat strange in the Swedish setting. The oldtime estates and guilds have been succeeded by interest organizations covering most groups and activities in daily Swedish life. In contrast to their American counterparts, these organizations participate not only as pressure and lobby groups to influence policymaking; they are formal members of the Royal Commissions that design policy alternatives and of the agencies implementing the selected policies. Organizational rather than individual participation is institutionalized in the policymaking process, and competence rather than citizenship is the capacity required of participants.[35]

Confronted with a choice between participation and performance, the Swedish policymaker will thus be inclined to prefer performance. The floor remarks by Göran Karlsson, a leading Social-Democratic spokesman, very clearly reflect this dominating order of preferences: "Think how the concept of due process can be used! If you have no other arguments left, it sounds terribly good to point to due process and the rule of law when you do not want to achieve something new. . . . I find it remarkable that one would like to disqualify the NEPB officials, who really have the competence in the field. . . . One wants results, not just talk."[36] The trust in the effectiveness and capacity of central bureaucracy was clearly evident in the committee report on the proposal to work out an environmental policy program in the Riksdag: "A more effective form of environmental planning . . . is the policy activity that the NEPB carries out as the central, leading agency in the environmental field, . . . which will form the basis for the more long-term activities of the environmental agencies."[37]

The strength of constitutional, institutional, and cultural considerations is seen also in the choice of patterns for public participation. The idea that individual citizens should be able to bring suits in courts to protect the public interest, and thereby bring about a judicial review of administrative action, remained a politically unthinkable alternative to all but some of the Liberals. Besides, what could the individual participant do to increase the effectiveness of pollution control that the agencies could not do better? The preference for administrative effectiveness and planned development over and above public participation comes across in the arguments of Herman Kling: "When it comes to large-scale pollution . . . the individual has especially limited capabilities to make a contribution. It is thus necessary for the government to have special agencies and special expertise for continuous control and monitoring of the pollution picture in different places and to take necesary action. Much of what was earlier the individual's own responsibility can now be transferred to the controlling agencies." As was shown in chapter 5, the minister believed that this would always be beneficial to the individual.[38]

We may thus conclude that political considerations indeed played a central role also in the Swedish policymakers' choice. Because of the features of the electoral and party systems, and because of the close links between parties and organized interests—and especially between the governing party, the administrative agencies, and the regulated interests—policymakers had strong incentives to select alternatives that would be acceptable to regulated interests and within the capacity of administrative agencies. Strategically, there was little incentive to try to capture the increasing environmental opinion by proposing escalatory alternatives. Considerations of political substance were also important; the majority of policymakers proposed alternatives well within the institutional structure of Swedish government. The choice was also consistent with dominating norms and beliefs in Swedish political culture; practicability and performance were sought before participation and due process.

The Hare and the Tortoise—Two Patterns of Choice

The analysis makes it clear that politics means more to policy content than environment and technology. The policymakers' perceptions of the air pollution picture and of the technological and socioeconomic consequences of different policy alternatives do not seem to have been consistent with the actual circumstances in each country. The analysis shows that the policymakers gave different weights to environmental, technological, and socioeconomic considerations because there were different sets of political incentives and disincentives at work, both in terms of political substance and political strategy. At every stage, considerations of the political costs and benefits of different policy alternatives played a decisive role in forming the policymakers' final choice. From what the analysis gives us reason to believe about the goals and preferences of policymakers, the patterns of choice seem consistent with rational assessments of the political costs and benefits provided by the different political contexts of choice.

The drastic American shift to health protection as the main policy objective was made because policymakers thought they could gain strategically from a tough line in response to widespread public opinion. Not subject to strong discipline within parties committed to long-term policy objectives or strong ideologies and not responsible for implementing the tough choices, American policymakers downplayed the socioeconomic and technological costs; the short-term strategic gains outweighed such costs. In terms of political substance, however, they did not escalate federal power beyond what would be constitutionally legitimate in terms of states' rights. And in the choice between performance and participation, considerations of political substance and political strategy both pointed in the same direction: increased public participation would be consistent with constitutional canon and cultural beliefs and would provide strategic gains in view of mounting public pressure for environmental quality.

The Swedish reliance on feasibility and practicability as policy criteria was continued because Swedish policymakers had no strategic incentives for drastic change. But it was also continued because the ruling Social-Democratic party was firmly committed to long-term policy objectives and an ideology stressing full employment, economic growth, and social welfare. The long-term consequences of clean air policy alternatives to the achievement of objectives in other policy areas were more important to the Social Democrats than short-term strategic gains. Politically responsible for policy implementation, the Social Democrats judged centralization and effective performance to be more important than participation. There were no substantial obstacles to centralization of policy powers in unitary Sweden. In strategic terms, there would be much to gain from effectively implementing a modest policy; the Social Democrats seem to have assumed that when the public saw the results, it would support the policy without actually participating in it.

We have discovered two distinct patterns of choice. However, we also seem to have found two distinct styles of choosing and selecting policies, in many ways resembling the old fable of the hare and the tortoise. Because of the peculiarities of the United States policymaking context, the American policymaker will be very aware of the reactions among his audience. The carrot that coaxes him forward is reelection and increased political status and power, all of which he can enjoy as a person rather than a representative of a party. In times of rapidly increasing political pressure from public opinion, the United States policymaker will thus be inclined to search for radical policy alternatives. In so doing, he will probably be less apprehensive about the problem of whether or not he will ever reach the goal and the glorious victory. In other words, it is more rewarding for him to speed up his action now than to match his speed and performance to the strength and resources objectively available to him. He has more to gain from pleasing the audience and making sure that those affected by his chasing down the road will accept and not try to obstruct the race than from carefully planning each stage of the race. Should stumblingblocks appear, the American legislator can always point out that they are due to circumstances outside his control, such as bad administrative performance despite generous resources.

In Sweden, the policymaker is somewhat more shielded from the immediate pressures of public opinion. Under the shell, the policymaker is presented with strong incentives to adjust the policy speed to the strength and resources available. There are also strong incentives to try to have all four legs—the cabinet, the Riksdag, the bureaucracy, and the regulated interests—moving at the same speed and in the same direction. What counts is not to set up a tremendous speed now, but to plan carefully each step along the road to ensure that the goal will be reached in due time. Not even very strong public pressure will tempt Swedish policymakers to increase speed and effort beyond capacity.

Our earlier analysis indicates that there is a close connection between these styles of choosing and the profiles of choice. The compressed and competitive style of choosing corresponds to a speculative augmentation of policy objectives and requirements beyond implementative capabilities. The continuous and consensual style corresponds to policy choices well within implementative capability. To say this, however, is to account for only half the race; it only raises questions of what happened after the start, the original policy choice.

It is especially interesting to speculate about American policy development. The enormous upsurge in public opinion made the 1970 context of choice rather extreme. Normally, also, the American process of choice includes the necessity for bargaining and compromise. In fact, textbooks would have us believe that this is the hallmark of American politics. But I think the fable could be used to enrich that picture. If it is true that "the hare jumped" at converging public environmental opinion in 1970, it follows that within the same institutional context, the policymakers will be attentive to converging public pressures in whatever field these may appear. To reap strategic benefits, the policymakers will leave the clean air policy and jump to adopting radical policies in the field of the day. What unfolds is a pattern of dramatic speculative augmentation of policy beyond implementative capability, followed by a long and complicated process of strategic retreats on policy objectives. With public pressure fading or gone, the policymakers will again be presented with a normal political context, in which there will be less incentive to reject the views of regulated interests, lobbying hard to have their version of what is feasible and possible adopted as government policy.[39]

What pattern of change could be expected in Sweden? The policy choice and the policy context of 1969 precisely fit the picture of policymaking given by textbooks on Swedish politics. The policymakers were under no pressure to come up with a final solution to the pollution problem. They deliberately labeled it a small step to be followed by others as the knowledge, resources, and necessity for further action became available or evident. What we can envisage, then, is the pattern of a tortoise moving slowly down the road, carefully assessing each step and perhaps incrementally changing its speed as it feels strong enough to do so. Could even a "shellshock" like the energy crisis around 1974 and the upsurge in public opinion following from that event induce the tortoise drastically to change its schedule? And what would the "changing of shells"—the installation of a new, bourgeois government in 1976—mean to policy change?

THE CHANGE

CHAPTER 7

The Hare Is Getting Tired

Retreating From Statutory Deadlines

We have pointed out that policymakers in a balance-of-power system with a weak party structure are faced with a particular profile of substantial and strategic constraints. Incumbents must keep in mind that the public views them as personal representatives and expects them to make or to oppose policies to suit the interests of the constituency. They thus have a strong strategic incentive to opt for policy alternatives that they perceive as favorable to, or favored by, their constituents. When a particular policy course is strongly favored by the public, the policymakers will find it advantageous to go along with these demands, rather than to argue that such a course is not practicable. Compounded further by the built-in competitiveness of the institutional context of policymaking, the situation may lead to a preference for the desirable rather than the feasible.

We have seen that this preference for the desirable guided United States policymakers in their enactment of a radically new clean air policy. In 1970, their assessment of constraints and opportunities led to the selection of a policy aimed at an immediate adjustment of present conditions to policy goals rather than at sequential adjustments of policy to increased implementative capabilities. They saw great political merit in an all-out effort to turn things right again. They found it strategically wise to associate themselves with absolute goals and specific achievement dates.

In effect, policymakers institutionalized a very particular wave of public opinion. In doing so, they made a successful implementation of the tough policy dependent on several substantial conditions, none of which might prevail or come true. The public might not continue to demand clean air in the face of the costs incurred from policy implementation. The polluters might not be willing to change or capable of changing their processes and products to comply with such an exacting policy. The social and economic conditions might change in such a way as to make continued development along the original policy line virtually impossible.

If the pendulum of public opinion were to swing away from clean air or if the goal achievement at the statutory date were found impossible despite vigorous bureaucratic implementation efforts, policymakers would be left with no other choice but to adjust policy to what they perceive to be "the new realities." In the United States case we might expect a gradual relaxation of the clean air policy, either in the form of relaxing the standards themselves or postponing the deadlines for goal achievement.

First, we would expect a lot of discussion and concern about the feasibility and practicability of the original policy in view of the technological developments and economic implications emerging from the implementation efforts. More specifically, we would expect discussion about the substantial feasibility of clean air policy in the face of its implications for energy supply and energy self-sufficiency.

Second, we would expect proposals for large-scale changes in clean air policy to make possible all-out efforts in other policy fields. When public opinion focused on the energy policy as a result of the 1973 oil embargo, we would expect policymakers to flock under this policy banner, trying to reap strategic benefits from demanding far-reaching adjustments in clean air policy to satisfy the public's demand for a secure and cheap energy supply.

The question is whether this pattern of change, and this style of changing, will prevail only when policymakers are faced with a turbulent and crisis-oriented set of substantial and strategic constraints. Can we expect large-scale adjustments in policy speed also when the oil crisis and the energy squeeze begin to fade away as the most important constraint on clean air policy choice?

Auto Emission Deadlines Suspended

As a result of the political judgment by Congress that dramatic legislative action was required to force the auto industry to develop a clean engine, the 1970 Clean Air Act Amendments directed the EPA administrator to set emission standards for 1975-76 that required a 90 percent reduction from the emissions allowable to 1970 model year vehicles. The amendments also established a context for future choice and change of auto emission standards and deadlines. The 1975 and 1976 standards could each be suspended for one year by the EPA administrator, provided he made findings that (1) a suspension was essential to the public interest or health and welfare; (2) all good faith efforts had been made by the applying manufacturer to meet the standards; and (3) the applying manufacturer had established that the technology was not available or had not been available for a sufficient time to permit compliance with the statutory deadlines. Congress also directed the EPA administrator to engage the National Academy of Sciences (NAS) to conduct a comprehensive investigation of the technological feasibility of the 1975–76 standards. The administrator could not grant a suspension if the NAS reports indicated that "technology, processes, or other alternatives are available to meet" the statutory standards.[1]

Thus, the development of the auto emission standards was geared to determining the availability of practicable and feasible pollution control technologies to achieve the statutory standards. What the law demanded was some immediate reduction of auto emissions in order to protect public health. This, however, implied difficult trade-offs. Might not a pronounced health-protection line commit the auto industry to a technology that, in retrospect, would turn out to be unsatisfactory? And would not such a commitment make it difficult in the future to change to a more satisfactory technology, with fewer negative consequences, such as increased maintenance needs, fuel and engine performance penalties, and the necessity for very elaborate inspection and regulatory systems? Finally, what would be the ramifications for employment and economic development of a pronounced health-protection line against the auto industry, traditionally the very backbone of the United States economy?

The final NAS report issued in February 1973 concluded that technological feasibility of the 1975 standards was likely but perhaps not attainable on the statutory schedule. The majority view suggested that it might be prudent for EPA to consider a one-year postponement of the 1975–76 standards to "provide the manufacturers an opportunity to consider and implement alternative, and quite possibly, more generally satisfactory technologies with which to attain the goals" of the 1970 amendments.[2]

On April 11, 1973, EPA Administrator William Ruckelshaus granted a one-year suspension of the 1975 carbon monoxide and hydrocarbon standards. At the same time, he set more stringent interim standards for California than for the rest of the country, so stringent that they would require the use of catalytic converters on all 1975 passenger cars in California. By this phasing in of catalysts he attempted "to minimize initial production problems and their potential impact on the public," at the same time that car manufacturers would gain production experience "preliminary to use of catalysts on all conventional engines during the 1976 model year."[3] That the administrator made his decision primarily on public interest grounds was made clear by his statement to the crowded news conference that he granted the postponement to avoid the potential societal disruptions that auto makers had threatened would result if emission control technology was required on all 1975 models, as directed by the 1970 act.[4]

The hearings that followed in the Muskie subcommittee dealt at length with the health and environmental effects of the postponement. While some Senators found the EPA decision realistic and proper, Muskie contended that the auto industry had not made a good faith effort to explore the feasibility of other solutions besides the catalytic converter. He scorned the American auto industry as the " 'can't do' guys of today" who refused to use now available and less expensive technological alternatives. "The goal of the committee continues to be clean, healthful air," he said, and urged the auto industries now asking for additional postponements to "maintain their efforts, in good faith, to meet the requirements of the law until such time as a change in law can be justified."[5]

What could justify a "pull back for a period of time in reference to the bracketing of dates"? The answer was clearly indicated by Senator Randolph in the April 1973 hearings: "No consideration of automotive technology and the use of cars can be complete without discussing their impact on the nation's energy supply. Currently anticipated technology is driving fuel consumption up sharply. This is occurring at a time when we face the prospect of widespread fuel shortage. . . . We cannot afford to resolve one serious national problem—automobile pollution—by increasing another—the fuel shortage."[6]

With the oil embargo in the fall of 1973, the mounting fuel shortage suddenly became a full-fledged energy crisis. The frustratingly long lines at the gas stations and the slightly hysterical mood of the nation's car owners put increasing pressure on policymakers to solve the gasoline shortage problem. Because of earlier claims that the catalysts necessitated to meet the statutory standards, as well as the interim standards, exacted stiff fuel penalties, there seemed to be a clear-cut trade-off between health and energy and economy. The problem was further complicated by the EPA findings that catalyst-equipped cars emitted sulfates, a hazardous pollutant. Furthermore, increasing the production of low-sulfur gasoline to overcome that health hazard might hurt total fuel production. Finally, the United States auto makers had to make their decisions on 1975 models early in 1974. At issue, then, was whether these circumstances justified compromising the earlier overriding concern for public health.

As could be predicted, policymakers responded both quickly and dramatically to the rising demands for solutions to the energy crisis. In the following eight months, several proposals and several decisions were made to delay even further the achievement of the goals for auto emission control.

On December 17, 1973, the Senate extended until 1977 the original 1975 deadline for carbon monoxide and hydrocarbon standards, and ratified the earlier EPA decision to postpone the 1976 nitrogen oxide standard until 1977. The bill was "no retreat," said Senator Randolph, merely "a stretching out of the time for arriving at the statutory standards." As such, Senator Baker argued, it was a "fair compromise between the continuing need for clean and healthful air and the need to improve the gas mileage of cars." Instead of a fuel penalty, the catalyst, if properly installed and maintained, might save fuel "anywhere from 13 to 18 percent."[7]

Action on that bill was not completed before adjournment. However, the early months of 1974 witnessed a race between the White House, the Senate, and the House to extend the auto emission deadlines as part of omnibus energy emergency legislation. On January 23, President Nixon asked Congress to move quickly to freeze auto emission standards for two more years to permit auto manufacturers to concentrate greater attention on improving fuel economy.[8]

In the latter part of February 1974, Congress adopted a conference report on the last controversial energy emergency bill. Among other things, the bill suspended the 1975–76 standards until 1977–78, with a possibility that the EPA administrator could grant another one-year suspension.[9] President Nixon vetoed the bill on March 6, and the Senate later failed to override the veto.[10]

To avoid a second veto, the House Committee on Interstate and Foreign Commerce split the energy emergency bill into two, one of which contained proposals to suspend the 1975–76 auto emission standards until 1977–78. Only five days after the release of the committee report on this Energy Supply and Environmental Coordination Act, on May 1, 1974, the House voted overwhelmingly in favor of a two-year suspension of auto emission standards. Two weeks later, the Senate adopted a substitute version of the bill.[11] A conference report was adopted by the Congress on June 11–12, 1974.[12] Ten days later, President Nixon signed the two-year suspension into law (see table 8).[13]

The House committee report on April 26 clearly reflected how differently clean air policy choices were now perceived by policymakers. The purpose of the bill was "to permit certain adjustment of environmental requirements, so that the nation's essential energy needs may be met."[14] But to Representative Wyman and others, this adjustment was not enough. Wyman thought it "unwise and unnecessary for us to be so enormously wasteful of energy in this country as to insist on having everyone in the country have an automobile that is equipped with expensive emission controls unless there is an honest-to-goodness, down-to-earth public necessity for this." Claiming that there was no realistic public health interest in having emission controls on cars in 90 percent of the nation's area, he once again proposed an amendment to lift emission control requirements off *all* cars outside a few specified

TABLE 8. Statutory Changes in U.S. Auto Emission Standards and Implementation Deadlines, 1973 to 1977

Proposals for Implementation of the Statutory Standard	Proposal Date	Implementation Deadlines	Grams of Pollutants	
Original 1970 Clean Air Act Amendments	1970	1975	HC	.41
		1975	CO	3.4
		1976	NO_x	.4
EPA administrator suspends deadlines	April 11, 1973	1976	HC	.41
		1976	CO	3.4
		1976	NO_x	.4
New deadlines signed into law by President Nixon	June 22, 1974	1977	HC	.41
		1977	CO	3.4
		1978	NO_x	.4
EPA administrator again suspends deadlines	March 5, 1975	1978	HC	.41
		1978	CO	3.4
		1978	NO_x	.4
New standards and deadlines signed into law by President Carter	August 7, 1977	1980	HC	.41
		1980	CO	7.0
		1981	CO	3.4[a]
		1982	CO	3.4[a]
		1981	NO_x	1.0[b]

HC = hydrocarbons; CO = carbon monoxides; NO_x = nitrogen oxides
a. EPA can waive the 3.4-gram CO standard if the agency decides two years in advance that the technology does not exist to meet the standard.
b. A single four-year waiver of up to 1.5 grams NO_x is authorized for innovative technology having potential air quality and fuel economy benefits.

air quality regions until 1977. Why on earth should people in rural areas "have to have these damnable octane octopuses that are guzzling up gasoline," exclaimed another gentleman.[15]

Representative Paul Rogers and others defended the bill as striking a necessary and reasonable balance between energy, environmental, and economic objectives. Rogers stressed that the catalyst technology would lead to gasoline savings of up to 25 percent, and that such technology would, in fact, protect public health. And don't forget, said Committee Chairman Staggers, that the "auto industry desperately wants this bill." If the bill did not pass, there would be many thousands out of work "within the next . . . two weeks." Whether or not this was a decisive argument, the House defeated the Wyman amendment by a very wide margin and adopted the committee's proposal.[16]

Additionally, the Senate was far from the tough, unrelenting mood of 1970. Muskie said limited changes at this time would be of value, because they would "provide the auto-makers with the certainty needed to proceed with the . . . 1976 automobiles," as well as a mechanism which will "balance [our] environmental goals with what we perceive to be the long-term energy needs of this country." To the critics who felt that health and health alone should dictate the standards, he pointed out that the health criterion had been retained. The only compromisable issue, he said, was how rapidly that should be achieved. Others were as conciliatory; Senator Randolph said the present measure was a reconciliation between extreme environmental and energy positions. The delays were not a retreat but rather a "realistic short-term response to the current energy situation." The Senate adopted the auto emission delays by voice vote.[17]

Muskie characterized the 1974 changes as a special measure to meet a special situation, "not intended to set precedents." Evidently, other policymakers did not share his views. From early 1975 through 1976 and 1977, there were repeated attacks on the original 1975–76 auto emission standards and deadlines. Supporting the auto industry's demands, the Ford administration in January 1975 proposed delaying imposition of the final auto emission standards until 1982.[18] On March 5, 1975, EPA Administrator Russel Train suspended until 1978 the implementation of two of the 1975–76 standards (see table 8). His sole reason for doing so was the alleged thirty-five-fold increase in sulfuric acid emissions—hazardous to health—from cars equipped with catalytic converters. At the same time, Train recommended that Congress delay implementation of the final standards until 1982. Train's position was, however, undercut by his own agency one month later, when another EPA report revealed that the sulfuric acid emissions dictating his March 5 decision had been grossly overestimated.[19]

Throughout 1975 and 1976, work on a major revision of the 1970 act continued in both houses of Congress. The Senate Public Works Committee voted 12 to 1 on February 5, 1976, to postpone two and relax one of the statutory auto emission standards (see table 9). In an individual view, Senator Muskie characterized the reported bill as reflecting "the minimum changes necessary to improve certain features of the Clean Air Act to accommodate perceived economic conditions."[20] On March 9, 1976, the House Interstate and Foreign Commerce Committee narrowly rejected the Dingell amendment, which essentially postponed the standards until 1982. The committee then accepted the "Brodhead compromise," which would impose the final standards in 1981–82. (For details, see table 9). Paul Rogers, who was becoming more and more the leading clean air spokesman in Congress, characterized the compromise as "a significant victory that keeps us from moving nowhere and indeed backwards."[21]

The floor debates clearly revealed that the single-minded efforts in

TABLE 9. Legislative Proposals Concerning U.S. Auto Emission Standards and Implementation Deadlines, Presented to Congress in 1976

Proposals for the Implementation of Auto Emission Standards	Proposal Date	Implementation Deadlines	Grams of Pollutants	
Senate Public Works Committee reports the Clean Air Act Amendments of 1976 (voted on February 5, 1976)	March 29, 1976	1979 1979 1980	HC CO NO_x	.41 3.4 1.0^a
H. R. 10498 to amend the 1970 Clean Air Act Amendments	Early 1976	1980 1980 1980	HC CO NO_x	.41 3.4 .4
Dingell amendment voted down 16 to 20 by the House Interstate and Foreign Commerce Committee	March 9, 1976	1982 1982 1982	HC CO NO_x	.41 3.4 b
Brodhead "compromise" adopted 23 to 20 by the House Interstate and Foreign Commerce Committee[c]	March 9, 1976	1980 1980 1981	HC CO NO_x	.41 3.4 .4
Gary Hart amendment defeated 30 to 61 on Senate floor; followed by adoption of the committee proposal	August 5, 1976	1979 1979 1979	HC CO NO_x	.41 3.4 .4
Waxman amendment defeated 75 to 313 on the House floor (demanded stricter interim standards up to 1981)	September 15, 1976	1981 1981 1981	HC CO NO_x	.41 3.4 .4
Dingell-Broyhill amendment adopted 224 to 169 on the House floor	September 15, 1976	1982 1982 1982	HC CO NO_x	.41 3.4 d
Conference report killed by a Senate filibuster on the eve of Congress adjournment	September 30, 1976	1979 1979 1982	HC CO NO_x	.41 3.4 1.0^a

HC = hydrocarbons; CO = carbon monoxide; NO_x = nitrogen oxides.
a. The original 0.4 NO_x standard was made a "research objective."
b. The final NO_x standard was to be decided through administrative action by the EPA.
c. Committee report filed May 15, 1976.
d. The final NO_x standard was to be decided through administrative action by the EPA, and auto manufacturers were given a chance to ask for further delays until 1985.

1970 to "maintain public confidence . . . in . . . both the progress and the results" of a *health*-related clean air program were the exception rather than the rule in this policy field. Muskie pointed out that as a result of the preceding "campaign to eliminate the regulatory and enforcement tools necessary" to reach the 1970 policy goals, the 1976 amendments reflected many conflicting pressures. They awarded "a modest extension of time to the auto industry to overcome technical problems—to deal with the fuel economy problem—and to recover from the economic troubles of 1974–75." Senators Baker and Buckley defended the committee's phasing-in strategy for new emission control technology as a balanced judgment, reflecting the "major tenor of these amendments to adjust the goals set by the 1970 act to reflect economic and technology limitations."[22]

At the same time, the committee spokesmen indicated that technology was not the real limitation on policy choice. So why grant a delay at all? Showing to the Senate a Volvo 1977 three-way catalyst capable of meeting the original 1975–76 standards as well as the Congress's fuel economy standards for 1980, Senator Gary Hart urged the Senate to continue the technology-push strategy by adopting his amendments and by rejecting the revised nitrogen oxide standard (see table 9).[23]

Interestingly enough, Muskie used strategic arguments to defend the revised nitrogen oxide standard. The 0.4 standard had been used by the auto industry "to drag its feet, to delay movement on other pollutants. . . . So we undertook to establish a legitimate target," said Muskie, one which could not be used by the automakers as an excuse for further delays. In 1976, it was by relaxing the standards that the automakers would be induced to speed up efforts to achieve the statutory standards! The Senate followed the committee's line of reasonable accommodation of conflicting interests and voted down the Hart amendments.[24]

Heavily attacked by both pro- and anti-environmentalists, House floor manager Paul Rogers was unable to save the Committee's so-called midcourse adjustments to the 1970 act, which were intended to "assure our Nation's continued economic growth and vitality." The sponsors of the pro-environmental amendments (see table 9) did not intend to "paralyze industry by applying unfairly burdensome requirements," but characterized the committee bill as "too great a capitulation to one of the most intense lobbying and propaganda campaigns the Congress has witnessed." Using the Volvo three-way catalyst as well as the recent recovery of the American auto industry, they argued that there was "no sound evidence to oppose [the Waxman] amendment, unless one is concerned about the profit margin from auto sales."[25]

Sponsors of the Dingell-Broyhill amendment (see table 9) characterized their proposal as more environmentally sound, energy efficient, consumer-oriented, and protective of jobs than the other two proposals. Dingell himself conjured up the specter of stringent auto emission requirements propelling unemployment and leading to further dependence on "the Arabs"; as he put it, "anyone concerned with the all-too-high unemployment level in our country today" must reject the overly stringent auto emission standards proposed by Rogers and Waxman.[26]

Evidently, the House preferred employment and economic recovery to environmental quality. First, it rejected the Waxman amendment. Then, the House adopted the Dingell-Broyhill amendment, thus rejecting the committee bill.[27] The conference agreement (see table 9) was killed by a Senate filibuster on the eve of adjournment for the 1976 election campaign.[28]

Paradoxically enough, this meant that instead of getting a further relaxation of standards, the auto manufacturers were left with 1978 as the final year for imposition of the 1975—76 statutory standards. The automakers immediately claimed that the 1978 models could not meet those standards. By default, both houses of Congress and the administration agreed to postpone tougher standards until 1979.

Angered by these events, Muskie vowed to offer no quick fix to the auto industry in 1977. Even if the final fix did not come very quickly, it nevertheless turned out to be more beneficial to the automakers. After seven months of intensive lobbying and bargaining and after hard-won conference compromise, the 1977 Clean Air Act Amendments became law of the land, postponing the original 1975-76 carbon monoxide and hydrocarbon standards until 1980-81, and establishing a relaxed nitrogen oxides standard to take effect in 1982.

If anything, 1977 was not the year for floor managers of clean air bills. In the House, Rogers was unable to save the auto standards section of the committee bill.[29] Against his assertion that the bill provided flexibility to consider energy and economy as well as technological availability to meet standards and deadlines, Dingell and Broyhill conjured up all sorts of disruptive effects on economic recovery, employment, and energy savings that would follow from adoption of the committee bill. Narrowly defeating a last-minute effort by the Carter administration to save the White House line on auto emissions (the Preyer amendment, see table 10), the House then adopted the Dingell-Broyhill amendment by a much wider margin than in 1976.[30]

In the Senate, Edmund Muskie told policymakers that they had arrived at the last frontier, and reminded them of the risk of further delays. Senators Riegle and Griffin, in opposition, argued forcefully that any faster movement on auto emission standards than the one envisaged

TABLE 10. Legislative Proposals Concerning U.S. Auto Emission Standards and Implementation Deadlines, Presented to Congress in 1977

Proposal for the Implementation of Auto Emission Standards	Proposal Date	Implementation Deadlines	Grams of Pollutants	
S 919, the Riegle (D, Michigan) and Griffin (R, Michigan) bill, supported by the auto industry	Early 1977	1980 1980 1982	HC CO NO_x	.41 9.0 1.0^a
HR 4444, the Dingell-Broyhill bill rejected in Committee by a tie vote, 21 to 21 (later adopted on the floor)	April 27, 1977	1980 1980 1982	HC CO NO_x	.41 9.0 1.0^a
Senate Public Works Committee reported the Clean Air Amendments of 1977 (S 252) by voice vote	May 10, 1977	1979 1979 1980	HC CO NO_x	.41 3.4 1.0^b
House Interstate and Foreign Commerce Committee reported the Clean Air Act Amendments of 1977 by voice vote	May 12, 1977	1979 1981 1981	HC CO NO_x	.41 3.4 1.0^c
Preyer amendment (backed by the Carter administration) rejected by the House 190 to 202	May 26, 1977	1980 1982 1982	HC CO NO_x	.41 3.4 1.0^c
Dingell-Broyhill amendment adopted by the House, 255 to 139	May 26, 1977	1980 1980 1982	HC CO NO_x	.41 9.0 1.0^a
Revised Riegle-Griffin amendment, (later changed on the floor by the Baker amendment)	June 9, 1977	1980 1980 1982	HC CO NO_x	.41 3.4
The Baker amendment to the Riegle-Griffin amendment, adopted by the Senate, 56 to 38	June 9, 1977	1980 1980 1980	HC CO NO_x	.41 3.4 1.0^d

HC = hydrocarbons; CO = carbon monoxide; NO_x = nitrogen oxides
a. EPA could waive the 1.0 NO_x standard and allow the 2.0 interim standard to continue "indefinitely."
b. The .4 NO_x standard was made a research objective.
c. EPA could decide in 1980 to reduce the NO_x standard to .4 from 1983 models. If so, all cars must meet the lower standard by 1986.
d. Cars with innovative engines, good fuel efficiency, or innovative emission controls must not meet the 1.0 standard until 1982.

in their revised amendment (see table 10) would damage fuel economy, reduce car sales and thus employment, and only marginally improve air quality. In an effort to stave off the Riegle-Griffin relaxation efforts, Howard Baker, now minority leader, offered an amendment which was a compromise between the committee bill and the Riegle-Griffin proposal. Baker was supported by Muskie, who said he "agreed to cosponsor it in the interest of bringing the matter to a head and resolving it finally." The Senate adopted the Baker amendment, thus rejecting the committee approach.[31]

The conference took place under difficult circumstances. Auto industry leaders threatened to stop production if Congress did not change the statutory standards before its vacation. If no agreement were reached, it meant that the 1978 models, due in the showrooms in September 1977, would not be in compliance with existing law. After a final seven-hour session ending at 2:20 A.M. on August 3, the conferees finally agreed on the auto emission standards section.[32]

The final compromise was passed by Congress the following day. On August 7 President Carter signed the 1977 Clean Air Act Amendments into law (see table 8). Two years after the automakers would have begun shipping the 1976 models meeting the original 1975–76 standards, the hare halted (temporarily?) its strategic retreat from objectives. That retreat had resulted in moving deadlines for the two 1975 auto standards to 1980–81, and in making the original 1976 nitrogen oxides standard a remote research objective for demonstration cars.

Adjusting Compliance Deadlines for Stationary Sources

As was shown in chapter 1, the 1970 legislation provided for two types of ambient air quality standards: primary standards to protect public health and secondary standards to protect public welfare. In April 1971, the EPA promulgated such standards for six classes of pollutants. The 1970 act also envisaged a strict timetable for achieving these air quality standards. Within nine months of EPA promulgation of the standards, states had to submit an implementation plan to the EPA, outlining how the ambient standards were to be achieved and maintained within the state. The EPA then had to approve or disapprove, wholly or partially, such implementation plans within four months. The plans had to provide for attainment of the primary ambient air quality standard within three years from the date of EPA approval. However, if "necessary technology or other alternatives are not available or will not be available soon enough," or if primary standards cannot be achieved within three years with "reasonably available alternative means," the EPA administrator could extend the three-year period for another two years. In practice, this meant that the primary ambient air quality standards were to be achieved by mid-1975 or, at the latest, mid-1977.[33]

Several policy means can be utilized to bring stationary sources into compliance with implementation plans and ambient standards. None, however, brings only beneficial consequences. First, one can regulate the number and size of polluting sources within any given area. But such an approach spells land-use planning—planned growth—and is not readily compatible with dominant beliefs in American political culture. Second, low-sulfur fuels can be used for industrial processes and energy production. However, this could lead to severe economic and social disruptions in a country with an increasingly competitive demand for energy. Third, one can require continuous control of what comes out of the smokestack, by having polluters install scrubbers or other control technologies. An objection to this method is that it puts new economic burdens on polluters and leads to a never-ending dispute over whether or not the particular technology is reasonably available or practicable. Fourth, one can require intermittent control systems, i.e., techniques that disperse and dilute pollutants by the use of tall stacks as well as a switch to low-sulfur fuels during unfavorable meteorological conditions. But this method might be insufficient to protect public health in the long run.[34]

The major threat to health from stationary sources is sulfur dioxide emissions. Many states chose to control such emissions by regulating the maximum sulfur content of the fuel allowed to be burned. All over the United States, plants preferred switching to cleaner fuels rather than installing control equipment, such as scrubbers. Even before the oil embargo, there were warnings by the U.S. Council on Environmental Quality that domestic supplies of low-sulfur oil and coal were inadequate to meet the demand established by state implementation plans. At the same time, the council stated that stack gas cleaning technology that would allow the burning of high-sulfur fuels would not be available to more than a small fraction of United States facilities until after 1975.[35]

The Arab boycott dramatically highlighted this so-called clean fuel deficit. As the deficit grew, and the price of low-sulfur oil surged, pressures to convert back to the plentiful but polluting high-sulfur domestic coal mounted. In effect, this meant that there was a growing demand to adjust Clean Air Act requirements to economic and energy realities.

In one way or another, such adjustments were part and parcel of all the 1973–74 proposals to enact energy emergency legislation. The late 1973 and early 1974 efforts to pass such legislation failed, but the Clean Air Act adjustments were finally adopted as the Energy Supply and Environmental Coordination Act of 1974.

The House committee report on April 26, 1974, maintained that although the embargo had been lifted, the potential for crises remained. To increase the nation's capacity to avert such crises, the committee proposed (1) to give EPA authority to grant short-term suspensions from any stationary source fuel or emission limitations until June 30, 1975, because of the unavailability of fuel necessary to meet original requirements; (2) to give FEA authority to order plants to convert to coal if they had the capability and necessary equipment to do so; and (3) to give EPA authority to suspend fuel and emission limitations for coal-converting plants until 1979. The reason given for this long-term suspension was that coal-converting plants unable to get low-sulfur coal had to be given as much time as necessary to install the continuous emission control equipment necessary to reduce sulfur emissions, to permit the orderly development of technology. Quite clearly, the committee wanted to adjust policy to technological practicability and economic feasibility.[36]

So did Muskie, but he was less inclined to compromise the health-related primary ambient standards. Therefore, his substitute version of the House bill provided (1) that no EPA short-term suspensions could be granted to plants located in areas where primary standards were already violated and (2) that no coal conversion could be ordered for plants located in areas where such conversion would lead to violation of ambient air quality standards.[37]

On the Senate floor, both Muskie and Randolph defended the bill. Muskie said that the legislation was not, under the threat of crisis, abandoning environmental goals. "It is in recognition of the need to continue energy conservation efforts that . . . we are trying to propose a mechanism which will balance [environmental goals] with what we perceive to be the long-term energy needs of this country," he said. At the same time that the bill continued to protect public health, it promoted coal conversion and thus also the important goal of energy self-sufficiency.[38]

Senator Randolph was even more outspoken on the need to adjust clean air requirements to energy policy goals. In recent years, he said, the American people had not done well in finding a suitable and equitable balance between energy and environment. Earlier clean air implementation schedules now appeared unduly optimistic and too inflexible to accommodate the realities of uncertain and inadequate energy supplies. These schedules had brought about a massive switch to imported low-sulfur oil. The embargo had revealed how vulnerable and unreliable this approach was in terms of both energy supply and environmental quality. The only option was energy self-sufficiency; "as recent events exhibit, the goal of energy self-sufficiency also will lead our nation to a more reliable clean air program."[39]

The only opposition to the proposed adjustments came from representatives of congested city constituencies. They argued that the bill was an opportunistic move to compromise the public health requirements of the original bill without any energy savings whatsoever. However, in the aftermath of the oil embargo, energy self-sufficiency was clearly preferred to public health. The stationary source parts of the bill sailed through both houses of Congress. The conference version followed the House approach in permitting short-term suspensions of air pollution standards through June 30, 1975, if clean fuels were unavailable. As for long-term suspensions for plants converted to coal, the conference followed Senate language and permitted suspensions until 1979 only if primary ambient air quality standards would not be violated. On June 22, 1974, these adjustments were signed into law as the Energy Supply and Environment Coordination Act (see table 11).[40]

Neither Muskie nor Randolph perceived this limited program as a final answer to the need for reconciliation of environmental and energy policies. It was only a first attempt to be initiated while the Congress continued to review the Clean Air Act and examined the need for broader authority to reduce dependency on foreign fuels.[41]

Indeed, the 1974 legislation only raised further questions concerning the relationship between energy self-sufficiency and air quality. Demand for domestic low-sulfur coal was already far beyond supply. Increased production of high-sulfur coal depended on the

TABLE 11. **Statutory Changes in U.S. Air Pollution Control Requirements for Stationary Sources, 1974 and 1977**

Compliance Year	Proposal
	1970 Clean Air Act
1975	Compliance with SIPs*
1977	Possible extension of SIP compliance
	1974 Energy Supply and Environment Coordination Act
1979	Compliance deadline for coal-converters
	1977 Clean Air Act Amendments
1981	Compliance deadline for coal-converters
1982	Compliance with SIPs in nonattainment areas
1983	Compliance deadline for existing nonferrous smelters
1986	Possible extension of coal-converter compliance deadline
1987	Compliance with SIPs in severly polluted cities
1988	Possible extension of compliance deadline for nonferrous smelters

* SIP = State Implementation Plan

availability and feasibility of stack gas cleaning technology. Coal producers were reluctant to open up new mines as long as they were not sure of a continued demand after 1979. Their reluctance was increased by the fact that their customers, the electric utilities, argued over and over again that stack gas cleaning technology was unavailable, infeasible, and a simple waste of the consumers' money. If the 1979 deadline were retained, that could mean another switch back to such low-sulfur fuels as residual oil and natural gas.

The "great scrubber debate" continued throughout most of 1974. On the Hill such leading policymakers as Randolph, Muskie, and Paul Rogers held that scrubber technology was available and that energy and clean air policy objectives could thus be reconciled. Within the administration, however, the EPA had to fight an uphill battle against the White House and the Federal Power Commission, who were not convinced of the feasibility of scrubber technology. Siding with coal producers and coal users, they wanted to allow intermittent and supplementary controls as a means of meeting clean air requirements without having to install scrubbers.[42]

During a press conference on November 27, 1974, EPA's Russell Train announced that the president's Energy Resources Council had reached a preliminary agreement to adopt a strategy of moving to permanent controls. As he unveiled the implications of the agreement, it became clear that it meant a full endorsement of a technology-adjustment approach to policy development. For the strategy to work, Congress would have to approve extensions for some power plants of as long as ten years, the main reason being that scrubber manufacturers could not possibly meet all orders for scrubbers until 1985.[43]

Consequently, the EPA endorsed the Ford administration proposals sent to Congress in January 1975 (see table 12). The Ford amendments (1) delayed until 1985 the mandatory use of scrubbers in coal-fired power plants in remote areas, (2) extended the final primary clean air deadlines up to 1987 for congested cities as well as for final particulate standards in regions where attainment is possible no sooner, and (3) extended compliance deadlines for coal-converting plants one more year, to 1980.[44]

When the Senate and House committee reports on the Clean Air Act revisions finally appeared in the spring of 1976, it became clear that Congress was prepared to adjust clean air policy not only to short-term crises but also to more permanent economic and technological realities. As Paul Rogers put it in an additional view of the House Committee report: "Some people have said that the 1970 act asked too much. . . . Maybe we asked for healthier, cleaner air too soon." But, he said, these amendments would make the necessary adjustment: "We have granted waivers, we have given extensions, we have written into the

TABLE 12. **Legislative Proposals to Change Air Pollution Control Requirements for Stationary Sources, Submitted to Congress in 1975 and 1976**

Compliance Year	Proposal
	1975 Ford Administration Proposals (not passed)
1980	Compliance deadline for coal-converters
1985	Intermittent controls for coal-fired plants in remote areas
1987	Deadline for reaching particulate standards in certain ACQRs
1987	Compliance with SIPs for congested cities
	1976 Clean Air Act Amendments (not passed)*
1979	Long-term suspensions of plant compliance schedules (1 and 3)
1982	Long-term suspensions of plant compliance schedules (2)
1980	Compliance deadline for coal-converters (1, 2, and 3)
1985	Also Compliance deadline for coal-converters (2)
1987	Compliance deadline for existing nonferrous smelters (2 and 3)
1981	Compliance deadline for plants using innovative technology (1 and 3)
1982	Compliance deadline for plants using innovative technology (2)

ACQR = Air Quality Control Region; SIP = State Implementation Program
*Compliance dates for 1976 Clean Air Act Amendments listed as set by 1) Senate committee proposal, 2) House committee proposal, 3) conference report

bill the flexibility to consider the economy, energy and technology available to meet goals and standards."[45] The Senate report succinctly stated the dilemma facing clean air policymakers six years after the enactment of the 1970 legislation. Of roughly 22,000 major emitting facilities, at least 3,500 did not comply with emission limitations or did not adhere to compliance schedules. Furthermore, many of these facilities were located in metropolitan areas, where ambient air quality standards were not likely to be attained in the immediate future. These metropolitan areas constituted the most logical and economical place for industrial expansion, but existing clean air legislation explicitly prohibited the construction of any facility likely to increase emissions in the area to prevent attainment of primary air quality standards! Last but not least, policymakers were under tremendous pressure from polluters to accept their contention that the technology deemed necessary to achieve the health-related ambient standards was not yet available and would not be practicable and feasible for a long time.[46]

The nation must have *both* clean air and a healthy economy.

Healthy air presupposed general use of continuous emission control equipment. A healthy economy necessitated an adjustment of pollution control demands to economic and technological capabilities. To get clean air *and* economic growth, both committees favored an approach of (1) extending compliance deadlines for stationary sources, (2) demanding continuous control equipment as necessary for final compliance, and (3) allowing industrial expansion in nonattainment areas if certain clean air requirements were met.

The two versions differed with regard to the length of compliance deadline extensions, with the Senate being more restrictive than the House (see table 11). On the other hand, both bills contained language explicitly demanding continuous pollution controls as a must for final compliance with control requirements. The Senate bill provided an incentive to industry to adopt innovations that will have wide application by extending compliance until 1981 for sources using new and more cost-effective control technologies. Particularly to encourage innovative technology that both reduced emissions and saved energy or reduced costs, the House bill extended until 1982 the final compliance deadline for plants using such technology.[47]

Both committees recognized that man cannot live by clean air alone. To permit continued industrial expansion in areas that had not yet attained the health-related primary ambient air quality standards, both bills required expanded or new sources to use the best available control technology. However, while the Senate bill allowed no increase but rather demanded a substantial decrease in total emissions after such industrial expansion, the House bill allowed for a 15 percent increase in total emissions until final compliance, in 1980. At that time, total emissions could not exceed the total amount of emissions before the new source located in the area.[48]

Senators Muskie and Buckley defended the emphasis on continuous controls and technology development as reflecting the philosophy of growth under an umbrella, imposing "reasonable controls as quickly as practicable [and] making staged improvements toward air quality." But others found the umbrella undersized. To lift the ceiling on economic growth contained in the present act and allow more weight to the overall social and economic welfare in deciding whether and where to permit growth, Senators Morgan and Allen proposed that new industrial plants should be allowed in nonattainment areas if "the social, economic, or environmental benefits of the new facility outweigh any benefits of preventing its construction."[49]

Indeed, this amounted to much more than adjusting the clean air policy to available means of implementation! It meant abandoning the earlier health-centered approach and entering the course originally selected in Sweden: a polluting activity may be permitted if it is of great importance for the economy or for the locality or if it otherwise serves the public interest.[50]

Quite understandably, Muskie reacted strongly: "If this amendment were to be adopted . . . the basic health protection purpose of the law would be null and void." But Committee Chairman Randolph was concerned about all of the potential obstacles to industrial growth in nonattainment areas. He said he would pursue the matter in conference, since the House bill was much more generous in these aspects. He also thought further compliance deadlines extensions might be necessary because of the anticipated shortages of clean fuels and appropriate control technologies.[51]

But there was little for Randolph to pursue in conference. The House first rejected an amendment that would have allowed the 15 percent increase in total emissions to last until 1982. It then adopted an amendment that deleted the whole section on industrial expansion in nonattainment areas. A last-minute effort to have the bill recommitted with instructions to allow for more stationary source variances was rejected.[52]

The conference agreement, which followed Senate language on compliance deadline extensions and adopted more stringent House demands for available technology for all new sources, was killed in a Senate filibuster as Congress adjourned for the 1976 election campaign.[53] A new showdown over the thorny issue of adjusting health-related clean air objectives to perceived economic and energy constraints could thus be expected in the 1977 session.

Under the 1970 act, several thousand stationary sources would become unlawful by mid-1977 because they did not meet that compliance deadline. In December 1976, the EPA issued regulations establishing a franchise system for expansion of polluting industries in nonattainment areas. New or expanded plants would be allowed only if (1) they used best available emissions-reducing technology, (2) their emissions were offset by reductions in emissions from existing sources, and (3) the owner's other plants were all in compliance or on a compliance schedule.

The 1977 committee proposals followed much of this pattern. They explicitly required continuous controls (i.e., best available control technology) as the only means of achieving and maintaining the health-related ambient air quality standards. On the other hand, economic and energy realities made it mandatory to allow some flexibilty in achieving full compliance through installation of costly control technologies.

Thus the Senate committee gave the states until 1979 to revise their implementation plans to ensure that emissions from stationary sources, including coal-converting plants, would be low enough to achieve ambient standards by 1982. Areas with severe mobile source pollution subject to transportation controls could be given a waiver until 1987. Even further relaxations and delays were contemplated. The committee first adopted but then—after heavy debate and opposition—relaxed an amendment by Senator Bentsen striking out the fixed compliance deadlines and requiring states only to make reasonable further progress toward primary ambient air quality standards.[54]

The House committee bill differed from the Senate bill with regard to some of the final compliance deadlines. Furthermore, it required states to make reasonable further progress toward attaining the air quality standards. The states were required to show equal emissions reductions every two years between 1977 and 1982, when final compliance would be required.[55]

Remaining differences with regard to stationary source regulations were easily settled in conference on July 27, 1977. States submitting revised implementation plans by 1980 were given until 1982 to achieve the primary ambient standards; the deadline was stretched out until 1987 for cities with severe oxidant and carbon monoxide problems. The twin principles of best available technology and equal offsets in emissions from existing sources would guide industrial expansion in nonattainment areas. States must show annual incremental reductions in total emissions as well as sufficient progress toward attaining primary standards. Coal-converting plants were given up to the end of 1980 to comply with emission control requirements, with the possibility of a second waiver up to January 1, 1986. Nonferrous smelters with severe sulfur dioxide problems were given up to 1988 to comply fully with all pollution control requirements.[56]

On the Senate floor, Muskie on June 8 warned all of the interests affected by the bill that no more "quick fixes" would be forthcoming. Ironically enough, he was right, although in a sense he probably did not intend in his warning. Under the 1977 Clean Air Act Amendments, signed into law by President Carter on August 7, 1977, clean air would not be as quickly fixed as Muskie had thought back in 1970. About ten weeks after all stationary sources would have come into compliance with the original 1970 standards, the deadlines for achieving those health-related air quality standards were delayed for five, and in many cases ten, years.[57]

A Strategic Retreat on Objectives

Between 1970 and 1977, the United States context of clean air policy choice underwent some dramatic changes. In 1973–74, the fuel shortage and the oil embargo drastically changed the socioeconomic conditions for the policymakers' substance calculi. Beginning in 1974, the economic recession compounded these changes and provided new constraints on clean air policy choice. As the implementation of the original clean air policy continued, several unforeseen physical-environmental and technological problems, such as the sulfate emissions from catalytic converters, began to appear, thus making the substance calculi more complicated and difficult than ever. Accordingly, environmental issues no longer received unqualified electoral support, and the now well-established environmental lobbyist groups had to fight an uphill battle against the traditionally strong industrial and utility lobbies.

All available evidence suggests that the United States clean air policy underwent dramatic changes that corresponded with these contextual changes. Auto emission deadlines were postponed for a total of six years, and the achievement of primary air quality standards through state implementation plans and source emission controls was set back for five to ten years.

What emerges is the picture of the hare, initially taking off with great leaps but gradually getting exhausted and therefore reducing its speed to fit its strength. In 1970, a majority of policymakers preferred alternatives that would adjust reality to policy goals. In 1974, as well as in 1976–77, a majority of policymakers preferred alternatives that would adjust policy goals to what they perceived as unyielding realities or more important national goals. In 1977, the 1970 majority in search of a strong clean air policy had become a small minority, consisting mostly of representatives from heavily polluted large metropolitan areas. In 1974 and 1976–77, the 1970 policy fathers engaged in heavy bargaining with affected socioeconomic interests to find compromises that would save as much as possible of the original policy and at the same time adjust policy to the pressing realities of the day. From 1974 onward, a considerable number of policymakers seemed inclined to make wholesale changes in clean air policy to promote what they perceived as more important economic and social objectives. In 1970, a majority opted for the *desirable*. In 1977, there was a majority in search of the *practicable* and *feasible*.

It is important to note, however, that no changes were made in the basic premises of the law. There have been only limited challenges to the underlying health protection premises of the 1970 act, said Muskie in 1976. He went on to cite several reports showing beyond doubt that the requirements of the 1970 act were justified to protect Americans from the present "15,000 excess deaths per year, 15 million days of restricted activity per year, and 7 million days spent in bed" among the 40 million people adversely affected by unhealthy air.[58] The House committee's 1976 report wanted to emphasize the predominant value of public health protection by adopting language that would prevent courts from restricting the EPA administrator's health-protective decisions and "assure that courts will not circumvent Congress' statement of priority for preventive health protection."[59]

What seemingly had changed was the policymakers' assessment of the technological possibilities and socioeconomic consequences of going through with the push approach adopted in 1970. In 1970, Muskie and others seemed to consider it easy for the American economy and for American industry to comply with clean air policy requirements. Cleaning up the air had to be easier than putting a man on the moon, they said, and pollution control costs would be more than offset by the billions of dollars gained through decreased medical and other costs of heavy pollution.

In 1976–77, policymakers such as Senators Muskie and Gary Hart, as well as EPA's Russell Train, still seemed convinced that technology was available and practicable. Muskie fervently argued that American industry had dragged its feet and not made good faith efforts to develop and/or install the technology needed to achieve the health-related statutory standards. On the other hand, Muskie and, to a much greater extent, Paul Rogers were willing to listen to arguments concerning the socioeconomic consequences of continuing the strictly health-related push for new and expensive technology. As we have seen throughout this chapter, a majority of policymakers wanted greater flexibility in the administration of the act and opportunity for growth of national industrial capability.[60]

It would be easy to conclude that the 1973 oil embargo and the economic recession beginning in 1974 were decisive in turning most policymakers toward support of drastic reductions in the speed of clean air policy implementation. At the consumer level, air pollution control requirements seemed to create costs and problems totally unwarranted by the pollution problem itself. In the words of John Quarles, former deputy administrator of EPA: "The [auto] standards had been tightened significantly for the 1973 model cars, and by the spring of 1973 word was out that the current models were giving a very poor showing on fuel economy. . . . because of their novelty and the publicity given to them, the emission controls bore the brunt of the blame for poor performance."[61] At the national level, pollution control requirements directly increased imports of low-sulfur fuels, making the nation more and more dependent on a steady flow of fuels from the Middle East. Clean air policies seemed to prevent expansion in dirty as well as pristine areas, the former a logical place for, and the second often in desperate need of, new industry.

On the other hand, the technological conditions and socioeconomic consequences related to clean air policy were not as clear-cut as many policymakers and citizens seemed ready to believe.

Paul Rogers pointed out that contrary to common belief, the catalytic converter would result in fuel *savings* of up to 20 percent. In 1976, when the fuel crisis had long since faded away, Senator Gary Hart and several members of the House pointed out that the Volvo three-way catalyst would not only meet the original emission deadlines by 1977, but also the fuel economy standards established for 1980! At the same time, Edmund Muskie presented figures showing that more than one million jobs were associated with pollution control programs in 1975. In that year, the GNP was 1.6 percent higher due to pollution control expenditures. Over the decade ending in 1982, "the Consumer Price Index will have risen only two-tenths of 1 percent because of pollution control expenditures," Muskie argued.[62]

In 1970, such arguments would have carried the day and led to the adoption of a tough pollution control policy. In 1976–77, however, they were easily overthrown by arguments favoring a delay in the achievement of clean air policy objectives. This was true despite the facts that the auto industry was recovering from the recession and that the acute fuel shortage had long since faded away.

What was decisive in tipping the balance was the policymakers' perceptions and assessments of the changes taking place in the strategic context of choice. The policymakers could easily perceive the sometimes subtle, but more often dramatic, shifts in public opinion and in the political clout of different pressure groups. In the words of John Quarles:

> The record of the three years from 1973 through 1975 reveals that the environmental movement had entered a new phase, a phase best understood in reference to the dynamics of policy formation in our government. The year 1973 witnessed a dissolution of the national consensus that had supported demands for environmental reform. This was in part the result of the difficulties in carrying out the environmental programs already established and in part the consequence of the energy crisis. The situation was aggravated a year later by the recession. . . . The upshot was a marked decline in citizen activity to promote environmental causes. Public opinion polls continued to show public interest in environmental goals, but it could more accurately be defined as approval than as support. What was missing was the intensity that had characterized the activity of former years. At the local level citizens were not mounting attacks on the remaining problems. The meetings and the mail fell off; every Congressman could feel the difference. And when the environmental movement lost the militancy of its grassroots leaders, it lost its political punch. The combination of these changes deprived the environmental leaders at the national level of the leverage they needed to achieve fundamental changes in national policy.[63]

During the Train confirmation hearings in August 1973, Muskie already seemed aware of what was coming. Up to now it has been easy to generate support for policy goals, he said. But that honeymoon period was over; "we are getting into a crunch period and it is going to be . . . difficult to avoid the pressures of those who would throw away much of what has been accomplished."[64]

Several studies indicate that the environmental movement reached its peak in 1970 and then fell off until 1974. The percentage expressing a great deal of concern over air pollution decreased from 60 percent in 1970 to 46 percent in 1974. Developments after 1974 are subject to some dispute. However, recent results indicate that environmental concern is at a permanently lower level than it was in the early 1970s.[65] Furthermore, polls reveal a change in priorities among the "Big E's": Energy, Environment, and Employment. While air and water pollution were mentioned by 4–6 percent of those asked about the most important problem facing the nation in 1970–72, they were not mentioned at all in similar 1974–75 surveys. Energy and employment were not mentioned in the 1970–72 polls, but got high percentages in both 1974 and 1975.[66] The percentage viewing the United States energy situation as serious never dropped below two-thirds between September 1973 and February 1977. Asked about important ways of increasing the United States energy supply, 56 percent of a February 1977 panel considered it important to relax auto emission standards to save gasoline. At the same time, 47 percent found it important to relax emission standards for power plants to allow them to burn higher sulfur fuels.[67]

When there is such strong public concern for a particular issue, the American policymaker is faced with strong strategic incentives to opt for what he perceives to be popular standpoints. He will find much merit in policy alternatives that seem to satisfy public demands. Since the election campaign is at most twenty months away for all the Representatives and for one-third of the Senators, it is tempting to agree with what seem the most popular standpoints on issues of public concern. It is a matter of not falling behind in the race for reelection, because constituents are closely watching their policymakers' voting records on key issues.

What happened in 1970 was that the policymakers joined the environmental bandwagon. They judged the environmental issue to be so popular in the electorate's mind that it would be strategically rewarding to vote for clean air. In 1973–74, the energy crisis created a strong public demand for radical energy policy action. According to Quarles, both the media and the public seemed ready to blame environmental regulations for much of the energy crisis.[68] Hence the policymakers' readiness to formulate and adopt an energy emergency legislation that would do away with much of the clean air policy of 1970. As can be seen from Quarles's account of the events, public opinion in rural areas was very strongly against most environmental requirements.

Hence the near success of the Wyman amendment on the House floor in December 1973; with thirty-five more votes, Wyman would have swept away all clean air requirements for automobiles sold outside certain metropolitan air quality control regions. In 1970, such a two-car-production policy was ridiculed in Congress.

The energy emergency legislation was vetoed not for environmental but for economic reasons. Hence Chairman Staggers of the House Committee on Interstate and Foreign Commerce could decide to go forward with the "essential parts of the comprehensive package of proposals on which there is substantial agreement," i.e., the clean-air-related sections. That substantial agreement united both Richard Nixon and Edmund Muskie, who otherwise had such different views on the appropriate content of the United States clean air policy. In 1976, policymakers still seemed to think that "the grass is greener on the other side of the fence." Thus, Muskie stated that "in time of recession and economic adversity, all you have to do is put the label of jobs and energy on an environmental proposal and the votes flock to your banner."[69]

Once the pendulum of public opinion began to swing away from environmental issues, Washington policymakers also faced a different strategic context with respect to pressure groups. With strong public pressure for environmental proposals, Washington environmental lobbyists could always be sure that the situation in the electoral arena gave extra strength to clean air lobbying. But as environmentalism leveled off—in part due to the energy crisis and the economic recession, in part due to the idea that clean air regulations were somehow to blame for these problems—the environmental lobby groups lost the political clout provided by strong electoral support. Instead, they now had to rely on their skills in traditional subgovernment and cloakroom policymaking, as well as on the substantive merits of pro-environmental arguments. At the same time, they now were faced with a competitive industrial lobby that not only controlled superior economic and information-gathering resources, but also could count on considerable public support.[70]

Business and trade-union lobbyists were very active during the 1976–77 process of changing clean air policy. Apart from the auto manufacturers and the utilities, the United Auto Workers played a crucial, indeed pivotal, role. Going along with the tough Muskie approach in 1970 and as late as 1973, the UAW in 1976 worked closely with Paul Rogers and supported the Brodhead compromise against the auto-industry-backed Dingell amendment. In 1977, however, the UAW sided with the auto manufacturers to have the auto emission controls eased. This alliance succeeded in weakening the standards. One policymaker lamented: "What are you going to do when you have Henry Ford and Leonard Woodcock on the same side?"[71]

Both the environmentalist and industry/union sides lobbied heavily during the 1977 round of policy change. One account of the House debates states that "during each roll call, the corridors leading to the House floor were chaotic, jammed to obstruction with gesticulating lobbyists. House members had to pick their way through a small forest of upturned and downturned thumbs, the traditional lobbying signal." In the Senate, Gary Hart wryly pointed out that "those lobbyists out there are lining the reception hall wall to wall, shoulder to shoulder." And both sides thought the lobbying had been successful. Industrialists considered the auto emission section as a gain, while environmentalists thought they had been successful on stationary source issues such as nonattainment and nondeterioration. As for administrative lobbying, it seems to have been better organized, and somewhat more successful, in the Senate.[72]

It is important to note the changes in the institutional context between 1970 and 1977. In 1970, there was no coherent environmental bureaucracy. But in 1974 and 1976–77, the EPA acted as a powerful agency whose commitment to specific policy proposals could be crucial in determining policy change. But EPA's strategic importance for Congressional policymakers is dependent on both EPA's own credibility and on EPA's backing from the rest of the executive. EPA's credibility with states and cities—and thus with Congressional representatives—was somewhat in doubt following the agency's efforts in 1972–73 to impose transportation controls and development controls on urban areas. States and cities were flabbergasted with what they considered unrealistic efforts to stop growth. The EPA decision in 1975 to delay for another year the auto emission deadlines because of the sulfate problem was almost immediately challenged by contradicting evidence produced within the agency itself. As we have already seen, the EPA also had to fight an uphill battle within the Nixon Administration concerning the timetables for achieving ambient air standards, as well as on the issue of continuous emission controls.[73]

There were also changes going on within the legislative arena. Committee memberships changed in 1973 and 1975 as a result of the preceding elections. The 1973 changes in the Muskie subcommittee were said to have made it difficult to resist pressures arising from urban traffic controls and concern for the energy supply. After he assumed the chairmanship of the Senate Budget Committee, Edmund Muskie became much less active in the area of environmental policy. Policy leadership slowly moved to Paul Rogers in the House, whose policymaking philosophy seemed to be heavily inclined towards incrementalism and compromise.[74]

The dynamics of the strategic context of choice thus provided clean air policymakers with a set of constraints and opportunities fundamentally different from the ones existing in 1970. The question was no longer one of escalating policy to satisfy a public demand for a stern response, but one of avoiding an onslaught against clean air policy resulting from a rapid escalation of energy policy to satisfy public demands in that field. To policymakers interested in saving clean air policy objectives *and* retaining their political positions, the strategically most advantageous option was to engage in the search for a compromise policy that, in view of the present circumstances, seemed feasible and practicable enough to attract a majority of policymakers.

This strategic judgment is reflected in both the style of changing and the pattern of policy change. Unlike the short and compressed period of policymaking in 1970, the final 1977 Clean Air Act Amendments were preceded by long and careful investigations and consultations. There were months of public hearings. Markup sessions witnessed a careful reading clause by clause to make sure policymakers knew what they were doing. Paul Rogers engaged in bargaining and compromises with the United Auto Workers, to "make sure the UAW was satisfied that the bill would not close down plants or increase unemployment"[75] Indeed, it was clear that things had gone back to normal; policymaking was proceeding by way of bargaining and negotiations within powerful subgovernments, so that policy alternatives capable of attracting a majority of the policymakers in the legislature could be formulated and accepted.[76]

The context of choice prevailing in 1976–77 provided strategic incentives for a pattern of change fundamentally different from the original pattern of choice. With the strong pressure for energy self-sufficiency and economic recovery, it no longer seemed strategically rewarding to hold on to stringent deadlines for goal achievement, regardless of whether or not capabilities to meet those deadlines existed or would be forthcoming. Leading policymakers engaged in a long and complicated series of moves to adjust clean air policy standards and deadlines to what they perceived to be the present and future capability of governmental agencies and affected interests. In so doing, they felt called upon to carefully balance a broader set of considerations than existed in 1970; the energy and economic consequences of clean air policy alternatives were allowed to play an important role as constraints on policy choice. This, it should be noted, happened despite the facts that technology *was* available in 1976–77, that the United States auto industry *was* successfully recovering from the recession, and that the energy situation *was* improving.

The end result was a pattern of change that closely resembled the one prevailing in Sweden throughout the lifespan of clean air policy: (1) a preoccupation with the feasibility and practicability of means rather than the desirability of ends, (2) incremental adjustments of policy objectives to present and future capabilities, and (3) a propensity to accommodate a wide range of conflicting private and public interests.

To some, it might be tempting to react by accusing policymakers of buckling down to pressures from certain interest groups. Some environmentalist policymakers did just that. Commenting on the 1976 Senate filibuster, Gary Hart said that "what this suggests to me is that a few companies in a handful of industries can control the great Senate of the United States."[77] And how far is the reaction of Muskie's environmental chief of staff, Leon Billings, from Muskie's 1970 order to Detroit to build a car with which people can live! Said Billings about the 1977 Amendments: "Firms like GM, suppliers like US Steel are so big they think they are above Congress and can force it to change. So far, they've been right."[78]

More important than passing normative judgments, however, is discussing whether or not the policymaking style and the policy content found here correspond to what could normally be expected to occur in the United States polity. In times of strong and unequivocal public pressure, the institutional characteristics of the United States electoral, party, and governmental systems provide policymakers with strong strategic incentives to engage in policy escalation, to move forward in great leaps like the hare. However, given the diversity of the American political culture, such unified demands are not likely to last very long. As soon as they fade away, or develop in separate directions, the very same characteristics that provided incentives for policy escalation now provide incentives for incrementalism, indeed make that policymaking style necessary. The normal situation is one where policymaking proceeds very slowly, with policymakers involved in intensive bargaining and making compromises and adjustments that they perceive as feasible and practicable as well as instrumental in forming the majority needed to carry the alternative through to enactment.

In the end, the hare may thus be found to move more slowly than the tortoise ever had since the race first began.

CHAPTER 8

The Tortoise Keeps Moving

Moving Forward Ever So Slowly

We have pointed out that policymakers in a parliamentary democracy must keep in mind a particular mix of substantial and strategic constraints. Parties in power must keep in mind that the public holds them responsible for the success or failure of policies. They thus have a strategic incentive to opt for policy alternatives that have a decent chance of getting implemented, given the resources available. But policymakers in the opposition are faced with the same constraints. They must be aware of the chance that they might become responsible for actually implementing their policy proposals. In strategic terms, there is a lot of competition among the parties, but it is limited by a denominator common to all actors, their search for the feasible rather than the desirable.

We have argued that this preference for feasibility guided Swedish policymakers in the area of air pollution control. Their assessment of substantial and strategic opportunities and constraints led to the selection of policy alternatives aimed at sequential adjustments to future conditions rather than at immediate adjustment of present conditions to policy goals. They saw great merit in flexibility. They did not want to associate themselves with absolute policy goals and specific achievement dates. They judged it substantially as well as strategically advantageous to retain the possibility of stronger action according as implementative resources become available.

But in doing so, policymakers institutionalized incrementalism even further. Clarification of policy objectives, as well as the search for and assessment of ways and means of implementing the policy, was left to the environmental bureaucracy. Indeed, what public agencies prefer most of all is planned development and controlled change. What an agency wants is a long-term program with resources clearly committed, and where envisaged changes are incremental enough not to upset administrative routines and agency performance.

In this effort, however, they are somewhat at odds with the politicians. Both the governing party and the opposition would like to retain flexibility and freedom of future action in the policy area. At the same time, however, they find themselves dependent on the agency for much of the information necessary for their substance calculi. Unless there is a dramatic change in public opinion on the policy issues, we can expect the following patterns of policy change.

First, there will be an all-encompassing emphasis on incrementalism, since all actors see advantages in small-scale changes geared toward practicability and feasibility. However, we might expect differences with respect to the time-frame envisaged in proposals for change. The politicians may be expected to hesitate to go along with agency-sponsored or maverick suggestions for long-term program commitments.

Secondly, we can expect to find a lot of discussion concerning the actual existence of possibilities to change policy objectives and to increase the levels of implementative performance. In other words, the policy debate will concentrate not so much on the *need* for change as on the *possibilities* for change. After all, the profile of incentives and constraints perceived by the policymakers will encourage them to try to reap strategic benefits from discussions of marginal differences in the economic feasibility and technological practicability of alternative policy changes.

The question is whether this pattern of change, and this style of changing, prevails also when policymakers face a quite different set of substantial constraints. Could we also expect only incremental changes in policy speed and direction when the oil crisis and the energy squeeze began to change the context of choice for air pollution control?

Establishing New Guidelines

The meeting at the Volvo foundries in Skövde in December 1969 and the subsequent promulgation of emission guidelines for stationary sources provided the steppingstone for implementing the selected policy alternative—the individual source control approach. In the following years, the responsible agencies were busy making decisions on control measures for all of the individual plants and installations covered by the 1969 Environment Protection Act.[1]

In 1969 the Riksdag specifically pointed out that emission guidelines were only a limited first step. It was assumed that the NEPB would continually evaluate the adequacy of the guidelines and propose new ones according as experiences and resources seemed to make changes necessary and possible. In the longer run, however, it was envisaged that "ambient air quality standards should be developed, mainly for the reason that circumstances may make them necessary."[2]

In August 1973 the NEPB issued new and revised emission guidelines for stationary sources. The revision was based on the experiences of the agencies in implementing the 1969 act. But it was also based on another, quite familiar, principle, the combined judgment of the NEPB and the regulated industries concerning "what is presently technically practicable within a realistic economic framework." The 1973 guidelines added several new branches and industrial activities to the list of sources subject to emission control. The guidelines were strengthened for several sources, but made less stringent for a few others.[3]

At the same time, the NEPB indicated a change in the emphasis of air pollution control activities. The change was necessitated by the successful implementation of the ISC approach: "By the mid-1970s, we will have completed the major clean-up of air polluting sources with the help of existing emission guidelines. Then we will also have the results of the supervisory control activities, which should provide more accurate values for emissions from different sources. At the same time, new emission control technologies are being developed, thanks not the least to United States developments. Economic rather than technical factors will put constraints on future action. This leads to a need for ambient air quality standards so that we can make priorities based on desired air qualities."[4]

There was no problem finding out what should be given priority. Air quality studies carried out since the early 1960s all pointed in the same direction: the lower the normal winter season temperature and the larger the number of inhabitants in the community, the higher was the normal winter season mean for the concentration of sulfur dioxide in ambient air. Many Swedish communities were found to have wintertime values well above those where adverse health effects might occur.[5]

Neither did there seem to be any difficulty regarding what standards should be adopted. An NEPB expert group studying health effects of air pollution stated that the World Health Organization (WHO) recommendations on air quality criteria issued in 1972 should be accepted as tools in the planning of air pollution control in Sweden. But they added that the values for sulfur dioxide and particulates should not exceed for any considerable time (a few months) the values given as annual means in the WHO report, the reason being the big difference between summer and winter concentrations of sulfur dioxide in the air masses above Swedish communities.[6]

In August 1976 the NEPB issued immediately applicable ambient air quality standards for maximum allowable concentrations of sulfur dioxide and particulates. The NEPB also established planning goals for ambient air quality to be reached within a ten-year period. Since the standards concerned winter season means rather than annual means, they would secure an air quality much better than that where, according to the WHO, respiratory dysfunctions could appear.[7]

After a decade of policy development, Sweden was thus finally prepared to make public health an integrated part of air pollution control. Typically enough, it was done with little political fanfare and with little reference to any urgent public need or any mounting public demand. It was the result of planned administrative action rather than of intensive legislative discussion and bargaining.

But what would it take to achieve the planning goals for air quality? The solutions seemed simple and straightforward. Since space heating was the most important pollution source, a lowering of the sulfur dioxide level could be achieved mainly in two ways: (1) by lowering the sulfur content in fuel oil and (2) by centralizing space heating in Swedish communities. The NEPB was optimistic, stating that with more district heating and good municipal planning for space heating, there should be no difficulty in reaching the planned goals.[8]

What seemed simple and straightforward to a single-purpose agency was, however, much more complicated for policymakers, since they had to make trade-offs between many different policy areas and political preferences. They had to find out whether there existed other alternatives with more favorable economic environmental consequences. Economically, the NEPB recommendations had to be assessed in terms of their consequences for the balance of payments. With no domestic oil whatsoever and with little domestic desulfurization capacity it could be extremely risky to base space heating on low-sulfur oil; would there be any, and at what price? Environmentally, the NEPB proposal was nudging the politicians to make more explicit trade-offs between clean air and a cheap but steady energy flow to keep Sweden going.

Protecting Sweden from Sulfur in Fuel Oils

The context of choice had become institutionalized in 1968. A cabinet ordinance issued in November that year vested the cabinet with the power to promulgate maximum allowable levels of sulfur content in fuel oil whenever the cabinet judged such decisions possible and/or necessary. At the same time, the NEPB was ordered to deliver annual reports and proposals concerning further reductions of sulfur content.

Reports from the NEPB in 1970 and 1971 clearly showed the seriousness of the sulfur dioxide problem. Ambient air quality levels regarded as safe by medical experts were exceeded in 40 percent of all urban areas where regular measurements were made. The NEPB argued that the best way of solving the problem was to increase the use of low-sulfur oil. The agency envisaged a strategy of differentiation. Maximum allowable sulfur content would be lowered at different points in time for different geographic areas and for different categories of users. As a first cut, the use of fuel oil with more than 1 percent sulfur should be prohibited in the metropolitan areas of Stockholm, Gothenburg, and Malmö. By a gradual widening of the 1 percent area and further construction of district heating systems, sulfur dioxide levels could be cut in half.[9]

The NEPB soon found out, however, that incrementalism must be planned if it is going to work. Just to prevent further *increases* in sulfur dioxide emissions, the use of low-sulfur oil would have to increase by about 2.2 million tons every year. Since naturally low-sulfur oil would be forthcoming in only half that amount, the method of enlarging the 1 percent area on the basis of available amounts of naturally low-sulfur oil would be totally inadequate.

The NEPB therefore proposed the immediate promulgation of a long-term program for lowering sulfur contents in fuel oil. By 1979, the 1 percent area should comprise all of Sweden. The long-term program was necessary to make possible a planned increase in domestic desulfurization capacity. Because of the long lead time of planning and constructing desulfurization plants, the program must be adopted now. Otherwise, there would be too little low-sulfur oil available later in the 1970s. The costs of committing Sweden to such a program would be very high, but they had to be paid if Sweden was to avoid serious threats to public health and environmental quality.[10]

The NEPB proposal and the principles for sulfur emission control were heavily attacked in the Riksdag. Questioning the Social-Democratic minister of agriculture, Ingemund Bengtsson, the Liberals argued that the present program was inadequate. Sulfur dioxide emissions had increased by almost one-fourth since the control program was initiated in 1968! And even with increased domestic desulfurization capacity, low-sulfur oil would be such a scarce commodity as to make future price hikes inevitable. The Liberals argued that the costs of the NEPB program could easily increase by five times the amount estimated by the agency. Liberals and Conservatives alike pointed to the promising stack gas cleaning technologies now available. The present control program should be replaced by a strategy of regulating stack gas emissions rather than what goes into the oil burners. That would allow scarce and expensive low-sulfur oil to be used for better purposes than just to heat homes and buildings.

In rebuttal, the minister pointed out that the Riksdag had vested sulfur regulation powers in the cabinet. The cabinet was thus responsible for launching a feasible and practicable program. In his opinion, the NEPB approach had to remain the focal point of the control program. Such alternatives as stack gas cleaning, better insulation of buildings, and construction of district heating systems in more and more urban areas could only be complements to the NEPB approach, at least for the 1970s. However, the cabinet would take only the first two-year part of the NEPB plan. For the rest of the 1970s there should be a flexible plan to allow for gradual adoption of new alternatives as they became practicable.[11] It is notable that ten days after this questioning, the Riksdag accepted without debate an agricultural committee report endorsing the NEPB control program.[12]

Conservative and Center Party bills in 1972-73 proposed that the sulfur content regulation should be replaced by regulations of stack gas emissions. The arguments were economic and technological; the existing program was too vulnerable to price hikes on oil and very promising stack gas cleaning technologies were being developed by Swedish firms.[13] The proposals were turned down by the agricultural committee. Stack gas cleaning was too expensive for all but the largest polluters. The smaller plants must be able to choose less expensive ways of decreasing sulfur dioxide emissions. The present program made exemptions from the 2.5 and 1 percent limits on sulfur content in fuel oil, provided that the plants had acceptable stack gas cleaning systems. The present approach could thus be preserved, and further NEPB and cabinet initiatives awaited.[14]

In 1973 the NEPB released reports showing that stack gas cleaning installations would be economically feasible only for the largest Swedish plants. The costs per ton of separated sulfur would be prohibitive for smaller plants. The NEPB recommended that the gradual widening of the 1 percent area should be kept as the major approach, especially since Swedish refineries were now planning for large expansions in their desulfurization capacities.[15]

The Energy "Crunch"

Then came the fall of 1973, and with it the oil embargo right at the beginning of the Swedish winter season. With oil representing about 90 percent of all energy sources used for heating purposes and with the price hike following the embargo, what would be the fate of the 1 percent approach to control sulfur dioxide pollution?

Seemingly, the tortoise kept its pace. Having temporarily halted the program for ten months after the oil embargo, the NEPB in late 1974 proposed that the 1 percent area should be enlarged to cover most of southwestern Sweden by January 1, 1976. The size of the regulated area would roughly correspond to the one envisaged in the NEPB's 1970 proposal.[16]

In promulgating the NEPB proposal on May 29, 1975, the Social-Democratic minister of agriculture, Svante Lundkvist, appointed a new

Royal Commission to investigate the sulfur problem and recommend alternatives for further action. On the one hand, mounting fuel oil costs made it necessary to give very careful attention to the feasibility of policy alternatives. On the other hand, the inflow and downfall of sulfur over southwestern Sweden, and the resulting acidification of soils, lakes, and forests in that area, indicated that the negative environmental and health effects were beginning to have tangible economic repercussions. Furthermore, there was now a new constraint on policy choice. A few weeks earlier, the Riksdag had decided on a new and comprehensive Swedish energy policy. The commission was directed specifically to look into the ramifications for energy policy of different sulfur control approaches.[17]

In its report of February 1976 the commission showed that sulfur dioxide emissions had increased from 300,000 tons in the early 1950s to 800,000 tons in 1974. If not counteracted, emissions would reach 1 million tons in 1985. Seventy percent of all emissions would come from the burning of fuel oil. The effects were already quite alarming. Soils and lakes in southwestern Sweden were heavily affected. If sulfur pollution were allowed to continue, forest growth and productivity—one of the backbones of Sweden's future economy—would be negatively affected.[18]

The Swedish policy could not be to eliminate sulfur emissions totally, because this would jeopardize such important objectives as economic growth and full employment. The only reasonable and feasible alternative was a gradual lowering of sulfur emissions to the levels prevailing in the early 1950s, which would prevent further environmental deterioration. Several means should be used: (1) regulation of sulfur content in fuel oils, (2) stack gas cleaning, and (3) lowering the consumption of fuel oils.

In comparing the first two alternatives, the commission did not hide the problems connected with sulfur content regulation. It was like gambling. There was at the moment a surplus supply of naturally low-sulfur oil in Europe. However, future government regulations of sulfur content might increase the demand, and thus widen the price difference between low-sulfur and high-sulfur oil. But Sweden should form its decision on the basis of normal price differences, according to which the higher price of low-sulfur oil corresponded to a cost per ton of separated sulfur that would be only 75 percent of the costs of stack gas cleaning when using high-sulfur oil. By using such examples, the commission quite unsurprisingly arrived at the same conclusions as the NEPB. For most oil-consuming plants and industries, the only feasible and practicable alternative was to use low-sulfur oil. Widening the 1 percent area to all of Sweden would reduce sulfur dioxide emissions by 40 percent. But this must be done gradually up to 1985 to prevent price hikes on low-sulfur oil. Besides, the acidification of soils and lakes was no urgent problem in northern Sweden.[19]

However, to lower sulfur emissions to the 1950 levels also necessitated a drastic lowering of Sweden's fuel oil consumption. By far the most important development here would be the construction of large-scale district heating systems that used the cooling water from nuclear power plants. If such systems were to be established for the metropolitan areas of Stockholm, Gothenburg, and Malmö, sulfur dioxide emissions would drop by 200,000 tons, almost 30 percent![20] The Commission's proposal is given at a glance in table 13.

It seems clear that what the commission wanted was gradual but steady progress. First of all, it wanted to reduce the vulnerability and economic risks connected with the reliance on the low-sulfur approach; after all, one could not be sure of a steady and cheap supply of low-sulfur oil. Second, the commission wanted an approach that would stimulate the development of more efficient and less expensive stack gas cleaning technologies. In the future, sulfur content regulation should thus be tied to both sulfur emissions *and* the energy content of the fuel oil. In this way, the law would be "neutral" with regard to whether fuel oil users preferred low-sulfur oil or stack gas cleaning. Together with other measures, this new approach would cut sulfur dioxide emissions by 25 percent between 1975 and 1985. By that time, it would cost half a billion Swedish crowns a year (see fig. 11).[21]

Was this a strategic retreat from the objectives outlined in the NEPB program of 1971? That program envisaged October 1979 as the target date for making all of Sweden a 1 percent area. Evidently it was not such a retreat. The 1971 program would have kept sulfur emissions at or below the 1970 levels of 700,000 tons. The commission's proposals meant that in 1979 sulfur emissions would be only about 650,000 tons. The only difference was that northern Sweden, with its less urgent sulfur emission problems, would not become part of the 1 percent area until 1984. In spite of the shock of the oil embargo, the tortoise was planning to move forward slowly but steadfastly.

Just a few weeks before the 1976 general elections, the Social-Democratic cabinet sent a sulfur regulation bill to the Riksdag. The minister of agriculture, Svante Lundkvist, said that sulfur dioxide emissions and sulfur downfall were the most serious of the nation's environmental problems. Despite the transnational character of the sulfur problem, Sweden must take strong unilateral action to bring down its own sulfur pollution. Sulfur regulations should concern all fossil fuels. The gradual widening of the 1 percent area should continue, and the NEPB should take strong action to bring down sulfur emissions from individual industrial sources.

However, action against sulfur in thin fuel oil must be harmonized

TABLE 13. The Royal Commission's 1976 Proposal Program to Regulate the Sulfur Content of Fuel Oil

Date	Type of Action
October 1, 1977	Prohibition in the whole nation of all import and sale of thin fuel oil and diesel oil with a sulfur content higher than 0.5 percent weight
October 1, 1977	Prohibition against burning fuel oil containing more than 1 percent weight of sulfur in the 15 southernmost of Sweden's 24 regions (except the Gotland region)
October 1, 1979	Prohibition in the whole nation of all import and sale of thin fuel oil and diesel oil with a sulfur content higher than 0.3 percent weight
October 1, 1981	Prohibition against burning fuel oil containing more than 1 percent weight of sulfur in the 19 southernmost of Sweden's 24 regions (except the Gotland region)
October 1, 1984	Prohibition against all import, sale, and burning of fuel oil containing more than 1 percent weight of sulfur in all 24 of Sweden's regions

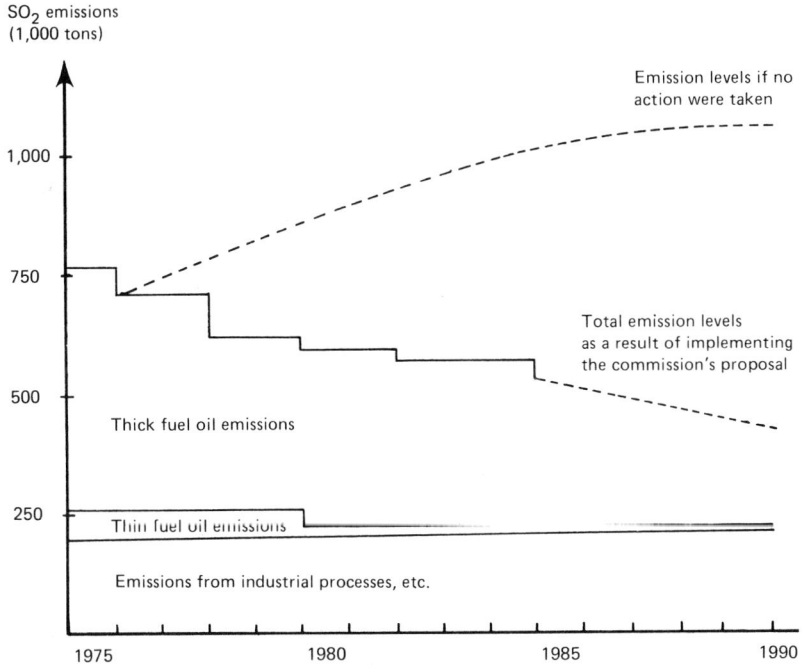

Fig. 11. Estimated effects on sulfur dioxide emission levels of the proposed Swedish fuel content 1976 program

with actions taken by the European Community. And like the Royal Commission, the minister did not want to commit himself to a control program for the period after 1980. Such a commitment could only be made in 1978, when the Riksdag was scheduled to make its final decision on Sweden's future energy policy. At that time, it would also be possible to decide on the issue of nuclear-power-based district heating systems for the metropolitan areas.[22]

A New Government Maintains Direction
When a newly elected Riksdag convened in October 1976, Sweden for the first time in forty-four years had a non-Social-Democratic government. Would the new Center-Liberal-Conservative cabinet and the new Riksdag majority continue to move forward with the sulfur pollution control program?

Several signs indicated that they would. In the cabinet declaration issued on the day of takeover, Prime Minister Thorbjörn Fälldin of the Center party stated that "emissions of sulfur and other air pollutants will be brought down." Center and Liberal private bills left over from the 1975–76 session contained several policy proposals quite similar to those proposed by the former Social-Democratic cabinet.[23] Now just an ordinary member of the agricultural committee, former Minister Svante Lundkvist could count on seeing his proposals accepted.

The committee was satisfied: "The proposals presented by the former government, and supported by the new government, are well balanced and will be instrumental in improving the air pollution situation."[24] However, the floor debate revealed the fundamental difference between Center party and Social-Democratic preferences concerning the trade-off between energy and pollution control policies. Having carried the 1976 election, thanks to its campaign against nuclear power, the Center party could not be expected to find it strategically advisable to support the idea of constructing nuclear-power-based district heating systems.

Former Minister Lundkvist jumped at this opportunity to associate the Center party with future increases in sulfur emissions. He pointed out that if the electric power and heat produced by the nuclear power plants decided upon by the former government were instead to be produced by fossil-fueled power stations, sulfur emissions would increase by 50 percent. He implied that if the Riksdag wanted to stop the fish kills and if it wished to restore soil and forest productivity and protect public health, then the Riksdag's final energy policy decision in 1978 must make nuclear power rather than fossil fuels the main Swedish energy source for the future.

The Center party chairman of the agricultural committee reiterated his party's view that the health and environmental risks related to nuclear power were of such a magnitude that the nuclear program must be derailed. But this would not lead to future increases in air pollution, since the oil used for power generation and heating purposes would, of course, be desulfurized. Others in the new governing coalition did not seem ready to accept the huge economic costs implied by the Center party standpoint. A Conservative spokesman lamented that already this limited sulfur control program put Swedish industry at a great disadvantage. He seemed to imply that an even heavier dependence on imported and costly desulfurized oil instead of domestically developed and produced nuclear power would impose unacceptable burdens on Swedish economy and Swedish industry.[25]

The Riksdag unanimously voted for the former government's proposal, and thus confirmed the incremental approach that had prevailed for almost a decade. However, the energy issue had already cast its shadow over the future course of clean air policy. The split over energy within the new coalition became visible only a few weeks after the takeover, implying that clean air policy might soon be placed in a much tougher and more profoundly strategic context of choice than it had ever been before.

Increasing Control of Auto Emissions

In December 1968 the Riksdag made exhaust emission control equipment mandatory for 1971 and later car models. From January 1970, the lead content in gasoline was subject to administrative regulation. Meanwhile, the Ministry of Transportation's Specialist Panel continued to work with the development of exhaust control technologies. Its final report and recommendations were expected in late 1970 or early 1971.

The environmentalists in the Liberal camp did not want to wait for the panel report. In a 1970 bill they called for an immediate adoption of more stringent emission standards for 1973 and later models, especially the standards for nitrogen oxides just recommended by the United States Department of HEW, insisting that such action was necessary to protect public health. The Riksdag committee agreed that action must not wait till the facts are all in, but added that standards should be tightened accordingly as technologies developed. The Liberal proposal was rejected on the grounds that the 1971 standards were not yet in force, and that some experience of these standards was necessary before any new standards were introduced.[26]

But the Liberals did not give up that easily. Pointing to the United States Clean Air Act Amendments adopted just a few weeks earlier, the Liberals' January 1971 bill strongly favored the technology-push approach. Technological practicability could no longer be considered a constraint on policy choice. Swedish car manufacturers had already stated that they could produce engines meeting the United States standards, and the Liberals said that "if they [the automakers] want to survive on the United States market, they will be forced to produce [such] engines." To give car manufacturers and administrative agencies reasonable lead time, the Liberals argued for an immediate Swedish adoption, in principle, of the United States 1975–76 auto emission standards.[27]

Then the specialist panel delivered its final report in April 1971. Auto emissions now accounted for 50 percent of Sweden's air pollution. And things would get worse. The number of cars would increase by 70 percent during the next decade. The control measures introduced for 1971 and later models would thus be completely neutralized as early as 1974.

The panel clearly preferred auto emission standards and installation of emission control equipment on all new cars as the most efficient means of preventing potentially hazardous increases in auto pollution. However necessary such a policy was to protect public health, it did imply some very serious economic and technological problems. First of all, would it be possible for Sweden alone to demand special control technologies, given the Swedish car export market's relatively marginal importance for many car manufacturers? Second, what would be the repercussions to the growing domestic car industry, now selling an increasing part of its production abroad? Third, what could Sweden possibly gain by awaiting coordinated European action? All earlier Swedish efforts had been in vain; the proposed ECE regulations had become so watered down that they could not possibly lead to acceptable air quality.

Fortunately for Sweden, there was now a way out of this deadlock. The newly adopted United States auto emission standards would force European and Japanese car manufacturers to develop emission control technologies if they were to survive in the United States market. Since almost every make competing on the United States market was also marketed in Sweden, the panel could see no meaning in having less stringent standards in Sweden than in the United States. Furthermore, the earlier differences between United States and ECE test driving patterns were vanishing. United States recommendations and standards would thus be immediately applicable to Sweden. The panel concluded: "The only realistic solution to the problem of strengthening the Swedish exhaust gas regulations seems, for the moment, to be an adaptation to the United States regulations."

But would not this be too great a leap forward for the tortoise? To be practicable and feasible one could not possibly follow the hare. To allow for the necessary lead time to develop adequate technologies and adjust production to the new regulations, the 1973 United States standards should be adopted for Sweden for the 1974–76 models (see table 14). Furthermore, the enormously stringent United States standards for 1975–76 models should not be adopted in Sweden until 1977.[28]

Still, the emission controls demanded for 1974–76 models presupposed a technology not hitherto tried on the European market. However, they were equivalent to the controls needed to meet the United States 1973 standards. By 1974 Swedish and European car manufacturers would thus have developed the technological capacity necessary to produce engines that met the standards. At the same time, the panel was not blind to the fact that international trade problems could occur. On balance, however, the scheduled adjustment to the stringent United States auto emission standards would "allow Sweden to obtain the advantages with respect to air quality [and to take] advantage of the new technical solutions . . . which must be developed if the United States requirements are to be met."[29] Indeed, the tortoise wanted the hare to run into and solve all the difficult problems.

TABLE 14. **1971 Proposal for a Program to Control Mobile Source Pollution**

Model Year	Maximum Emissions (gm/km)	Estimated Reductions in Car Emissions (%)	Lead Content in Gasoline (gm/l)	Comments
1973			0.4	Proposal made by the Poisons and Pesticides Board
1976	$CO = 24$ $HC = 2.1$ $NO_X = 1.9$	$CO = 60$ $HC = 70$ $NO_X = 45$		Equivalent to U.S. standards for light duty vehicles of 1973 and later models
1977		$CO = 80$ $HC = 80$ $NO_X = 90$	"Lead-free gasoline must be available"	Application "in principle" of the U.S. 1975/76 model standards; however, application in practice must be preceded by thorough investigations

With that, she would save time and strength and reach her goal more easily.

Reporting on the Liberal bill in October 1971, the Committee on Agriculture pointed out that the panel's proposals satisfied most of the Liberal demands. The only difference was the timetable for action. The Liberal timetable was not economically feasible; such swift unilateral Swedish action could lead to repercussions for Sweden's international trade. The cabinet should be given time to try to convince other European governments of the merits of coordinated European adoption of the United States standards.[30]

But not even the tortoise could stay motionless very long; the air pollution picture necessitated movement. In a May 1972 report on new Liberal demands to adopt the American timetable, the agricultural committee informed the Riksdag that the cabinet was busy trying to bring about concerted European action on the 1973–74 car models. If such action did not come about, the cabinet would be ready to "contemplate such strengthened regulations that are consistent with the demands for a satisfactory control of air pollution."[31]

In October 1972 the cabinet showed that this was not just empty rhetoric. It promulgated a new Auto Emission Control Ordinance, based on the 1971 report of the specialist panel. As one could expect, however, the cabinet did not commit itself to the panel's timetable. The new ordinance made the United States 1973 standards applicable only to 1976 and later models. There was no mentioning whatsoever of a possible adoption of the tough United States 1975–76 standards for 1977 and later models. The politicians wanted to play it safe; by 1976 one could be sure that the necessary technology would be at hand, and the economic costs would not deter prospective car buyers.

On the other hand, the new ordinance meant that to improve air quality, Sweden had put itself in a position between the stronger United States standards and the weaker European ECE regulations. Commented the NEPB: "This policy is not without problems. Among other things, it intrudes on the desired harmonization of European action and thus influences the car trade between Sweden and the Common Market. Auto emission control is thus the first area where environmental policy goals have been directly traded off against those of increased export and free trade."[32]

The Context Changes

The context of policy choice changed considerably during 1973. Earlier, the Liberals had fought alone for an earlier adoption of more stringent auto emission standards. Now the Conservatives and the Center party rallied to their support. Furthermore, the general election in September resulted in a tie between the traditional blocs in the Swedish Riksdag.

The Social Democrats could thus end up being defeated by lottery on the issue of auto emission control.[33] Finally, the oil embargo put new constraints on policy change. Stronger emission standards and sophisticated exhaust control technologies had the undesired effect of increasing gasoline consumption. Would it not be necessary to contemplate a redirection of the regulations, away from exhaust control technologies and toward new, less polluting fuels and engines?

The Liberal 1974 bill wanted *both* more stringent standards *and* governmental incentives for developing new fuels and new engines. In the majority view of the agricultural committee, the Social Democrats challenged the Liberal view: "The uncertainty about Sweden's future supply of motor fuels caused by the present oil crisis makes it extremely difficult to propose a further strengthening of the auto emission standards at this point." The Social Democrats countered the bourgeois parties' minority view that preparations for such a move must anyhow begin here and now by pointing to recent United States developments. The passage of the Energy Supply and Environment Coordination Act meant a two- or three-year delay in implementing the stringent 1975–76 standards. Under such circumstances, Sweden should not try to take the lead. The roll call resulted in a tie between the two views, but the lottery favored the minority view.[34]

The Social-Democratic cabinet made no effort to fulfill the bourgeois opposition's demands. But the 1974 lottery victory implied that the opposition might engage in swifter action against auto emissions as soon as they would get an opportunity to form a coalition government. That opportunity came with their victory in the 1976 general elections.

Now in opposition, the Social Democrats presented a party-sponsored bill in January 1977, calling for preparatory planning and action to strengthen the auto emission control program within one year. They said that available scientific evidence concerning the relationship between auto pollution and adverse health and environmental effects was strong enough to warrant immediate action. Evidently, they were not satisfied with their own earlier efforts to bring down auto pollution; the increasing number of cars outnumbered the control efforts. In fact, the percentage of total air pollution stemming from mobile sources had increased from 45 percent in 1969 to almost 55 percent seven years later. Within one year, the new cabinet should begin to implement a control program involving (1) more effective performance tests and controls of emission control equipment on new and used vehicles, (2) ambient air quality standards for auto emission pollutants, (3) systems for traffic planning and transportation controls, and (4) lowered lead content in gasoline.[35]

Since most of these proposals resembled earlier demands from the parties that now formed the new coalition government, one could assume that they would be easily accepted. On the other hand, the view of Swedish policymaking and policy choice taken earlier in this book might make us believe that once in power, the bourgeois parties would be as cautious in staying within the perceived limits of practicability and feasibility as the Social Democrats had ever been.

That is exactly what happened. The agricultural committee's bourgeois majority admitted that auto pollution was increasing, but asserted that the new cabinet was determined to come to grips with the problem. Thus, the Conservative minister of transportation had just appointed a Royal Commission to come up with proposals for a more effective system of testing and controlling the performance of emission control equipment on new and used vehicles. The Center party minister of agriculture was just about to appoint another commission. That commission would investigate the relationship between auto emissions, public health, and environmental quality, to find out whether and to what extent available evidence justified a further strengthening of the present auto emission standards. All further action must await the reports of these Royal Commissions.[36]

No, said the Social Democrats; there is enough scientific evidence to warrant immediate action. Their proposal was well founded. It represented the policy they had planned to introduce if in power after the 1976 elections. Earlier, the bourgeois parties, and especially the Center party, had excelled in presenting themselves as champions for the environment. If they wanted to prove their environmentalism, if they wanted continuity in air pollution control, it would serve them well to adopt our proposals, said Svante Lundkvist, the former minister of agriculture.[37]

The next day, the Center party minister of agriculture, Anders Dahlgren, presented the new commission's directives. They made it clear that the new cabinet was not inclined to take any immediate action to strengthen Swedish auto emission standards. The commission was expected to (1) assemble all available knowledge on the health and environmental effects of mobile source pollutants, (2) describe the latest domestic and foreign developments in control technologies, (3) evaluate the impact of earlier control measures, (4) evaluate promulgated and planned foreign auto emission standards, and (5) consider more stringent Swedish standards in view of the desirability of adjusting these standards to those prevailing in other countries. Economic feasibility must prevail as a main criterion: "Health and environmental benefits following from proposed action must of course be weighed against the economic costs of such action . . . the Commission's proposals must be guided by a long-term perspective."[38] According to NEPB experts, the Royal Commission will not present any report until the beginning of the 1980s.[39] Evidently, the tortoise did not change her speed just because she had changed shells.

In their 1977 bill, the Social Democrats argued for swift action to lower the maximum allowable lead content in gasoline from 0.4 grams to 0.15 grams per liter. West German experience showed that such a lowering was economically feasible; gasoline prices would increase by no more than 1 percent. The benefits for environmental quality and public health were so enormous that a decision should be taken at once. There were no legal or administrative constraints. The law vested the Product Control Board with the power to regulate lead content whenever the board found such action necessary and/or possible.[40]

If anything, incrementalism was characteristic of earlier choices in this field. An administrative ordinance lowered the maximum allowable lead content to 0.7 grams per liter of gasoline, effective January 1, 1970. The ordinance was based on the results of negotiations between the responsible agency and the gasoline producers. It was assumed that further lowerings of lead content would be based on the results of future negotiations.

As usual, the environmentalist Liberals were unhappy with this gradualistic approach. They cried that the terrible threat to human beings from lead exhausts necessitated swift action. Furthermore, there was technological capacity available for producing low-leaded or lead-free gasoline, which made drastic action practicable. They insisted that drastic measures were economically feasible; gasoline prices would increase by less than 1 percent. The maximum allowable lead content should be lowered to 0.4 grams per liter immediately and to 0.15 grams by 1976.[41]

The Riksdag's majority rejected these speed-up proposals on economic and technological grounds. The existing car population was not equipped with engines that could run on low-lead or lead-free gasoline. One must thus await technological breakthroughs that could make the switch economically feasible. Furthermore, the initiative should come from the administrative agency. The results of the negotiations between the agency and the gasoline companies must be allowed to determine the timetable for further regulatory action.[42]

In 1972 these negotiations resulted in a new ordinance, lowering the maximum allowable lead content to 0.4 grams per liter from January 1, 1973. Then nothing happened until the Social Democrats' 1977 bill. The agricultural committee reported that in April 1977 the Product Control Board established an action program for the further lowering of lead content. A formal decision on the dates for such lowerings could be expected by the end of the year. In December 1977 the Product Control Board recommended that the cabinet lower the allowable lead content level to 0.15 grams per liter, effective January 1, 1980.[43]

Approaching Objectives Step by Step

The years 1970–77 witnessed a series of dramatic changes in the substance, preference, and strategy contexts of clean air policy choice. The 1973–74 oil embargo drastically changed the substance conditions relevant to the continuation or change of existing clean air policy alternatives. The gradual development of a new, comprehensive Swedish energy policy implied that crucial trade-offs might have to be made between preferences for clean air and for a low-risk, steady energy supply. the 1976 elections changed the strategic picture by placing in power the Center party and the Liberals, both of whom had tried hard to present themselves as the Swedish environmentalist parties *par préférence*.

Despite these changes, all available evidence suggests that the Swedish clean air policy has moved forward, albeit in an incrementalist fashion. The sulfur regulation program has moved forward slowly but steadfastly, despite the shock waves of the oil embargo. By the end of the 1970s, the 1 percent area approach will even be somewhat ahead of the objectives outlined by the NEPB in its 1970–71 proposals. And despite the possible repercussions for international trade, as well as the setbacks in international negotiations, Swedish auto emission standards were gradually strengthened over and above those established for the rest of Europe.

What emerges is the picture of the tortoise, moving along with only slight modifications in speed and direction, seemingly shielded from the dramatic events at the roadside. With the exception of some individual Liberal Riksdag members, the policymakers seemed little inclined to make recommendations for dramatic policy changes. And even when doing so, the Liberals felt obliged to present their proposals as being well within the range of practicability and feasibility. The majority of policymakers seemed to prefer policy changes for which they could also take responsibility at the stage of implementation.

Indeed, adjusting policy objectives to available resources and foreseeable capabilities seems to have been the approach favored by all policymakers. Speaking against the opposition's proposal to strengthen the Swedish auto emission standards, and commenting upon the recent suspension of the United States 1975–76 standards, one Social Democrat said in 1974 that "I think we have been more realistic here in Sweden by recommending a gradual long-term strengthening of the auto emission standards, beginning with the 1976 models. . . . In this way we get more time to adjust to the standards we want to have. . . . I consider it much more important that the existing standards are adhered to than to promulgate new standards which could not be implemented."[44] When in power, the Center party people did not seem less inclined to adhere to feasibility as the main policy criterion: "We have also expressed far reaching aspirations and presented ambitious proposals, which we hope to be able to implement as soon as we get the resources."[45] Both blocs, when in power, seemed equally ready to adjust to what they perceived as technological and socioeconomic constraints. They all seemed to prefer incremental (but clearly feasible) changes to speculative (but perhaps infeasible) redirections of policy.

This is all the more intriguing in view of the unmistakable increase in the number of references to public health as a major criterion for future changes in clean air policy. The steady flow of research reports showing the adverse health and environmental effects of air pollutants was slowly but steadily perceived by policymakers as necessitating a certain restructuring of the preferences guiding clean air policy changes. In the final analysis, however, the majority of policymakers did not allow this preference for public health to tip the balance in favor of major redirections of the clean air policy. On balance, not even the adoption of auto emission standards more stringent than those for the rest of Europe was a break with the traditional pattern; public health was paid tribute to, but in no way did policymakers politicize clean air in the sense of allowing it to supersede, or intrude upon, such central values and objectives as economic growth and full employment.

Seen from a strategic viewpoint, this ordering of preferences, and the pattern of change following from it, is easily explained. At the time, environmental opinion existed as a potential strategic reservoir, but not as a coherent pressure group demanding specific changes in air quality or other environmental policies. In fact, just about 1 percent of those surveyed before the 1973 and 1976 elections spontaneously mentioned environmental policy as something they specifically liked or disliked in the parties' records and platforms. The number of voters spontaneously mentioning such policies as employment, taxation, and energy, was much larger.[46]

Evidently, the policymakers could count on these issues having a much greater strategic impact than issues of air quality. At the same time as it paid off to be pro-environmental in a general way, the parties had a wide latitude of choice when they went about changing the clean air policy. With such a general but vague environmental opinion, efforts to make political capital from environmental quality might lead to only marginal effects on votes and parliamentary strength. Neither would policymakers be strategically punished for suboptimization and incrementalism when changing clean air policy. On the other hand, efforts to promote clean air objectives at the expense of objectives in policy areas closely watched by the voters could lead to large cuts in votes and parliamentary strength.

To use a favorite expression of many Swedish policymakers, there was, from a strategic point of view, no "cow on the ice" in the process of clean air policy change. But woe to the policymaker who dared to propose clean air policy changes that could drive such sacred cows as Full Employment, Economic Stability, and the Tax System out on that ice! A majority of the policymakers judged it strategically wise to adjust clean air policy changes to the goals of other policies rather than to put it over and above well-established, popular policies, since the strategic effects of such a politicization were uncertain or possibly negative.

The fate of the efforts to switch from environmental regulation to environmental taxation vividly illustrates this point. Throughout the early 1970s, the Liberals argued for the introduction of a system of environmental taxes, fees, and duties. Fuel taxation should be changed to favor low-lead or lead-free gasoline or to favor alternative, nonpolluting fuels, such as liquefied petroleum gas (LPG). Vehicle taxation should be changed to favor unconventionally powered vehicles or to favor small cars over gas guzzlers.

These policymakers argued that economic incentives provide a much more effective and powerful means for achieving environmental quality and public health than administrative regulations ever could. Taxing fuels according to their polluting effects would provide direct signals to individual car owners to switch to cleaner fuels. Putting fees on polluters in relation to the amount of emitted pollutants would provide them with incentives "to reduce pollution in the least expensive way. . . . A system of fees provides a better feedback to administrative agencies who will get exact information about pollution amounts." Using the price mechanism would stimulate "new technology, new production methods, new goods with less harmful environmental effects. With regulations and exemptions one seldom provides incentives to those who reduce pollution in the fastest and least expensive way," argued the Liberals. They wanted to introduce a system of taxes, fees, and other economic incentives based on the criteria of public health and environmental quality.[47]

But these policymakers soon found out that the principles of taxation systems are very much like sacred cows. They are defended without much consideration as to whether yesterday's rationale is today's rationality. Committee and Riksdag majorities rejected all of the proposals, arguing that "there is no reason to use taxation measures as a substitute for prohibitive regulation in the field of environmental control." The fuel and vehicle taxes were based on the principle that each vehicle has a highway cost responsibility, and the taxes were specially destined to the highway tax funds. If one tampered with this principle by favoring LPG over gasoline, it meant that an LPG-powered car would not fulfill this responsibility; it would pay less toward the highway funds than a gasoline-powered car with the same weight and the same mileage. To this individual injustice would be added the collective effects. Tax incentives would increase the number of LPG-powered cars, and thus lower the total sum of tax money going to the highway funds. To maintain highway standards, the government might have to increase the LPG taxes. Many LPG car owners would then feel deceived by the government. In 1974 the committee concluded that "however important the considerations of environmental policy may be, they must not be allowed to distort the well-established principle that motor vehicles shall pay their part of the costs for highway construction and maintenance."[48]

By 1974 the proponents of environmental taxation had a new argument. The oil crisis made it necessary to differentiate taxes not only in relation to pollution effects but also in relation to fuel consumption. In a minority view, the bourgeois block stated that time was now ripe for a "revision of the Highway Traffic Taxation Ordinance . . . to stimulate the use of energy-saving and environmentally harmless vehicles." The cow was not sacred any more: "The principal issue today is not the size of the road traffic tax but its approach and objective." Are the Social Democrats "prepared to use it to save energy and achieve cleaner air?" one Liberal asked. The Social Democrats rebutted that "the environmental issue should not be discussed in connection with taxation." The roll call was a tie, but the lottery favored the bourgeois minority view.[49] To date, however, the new bourgeois government has presented no plans for an environmentalist revision of the tax system.

For the policymakers, notably the environmentalist Liberals, who perceived the physical-environmental and socioeconomic conditions as necessitating and making possible radical changes in clean air policy, the political context thus provided formidable constraints. In terms of political culture, the majority of policymakers and voters seemed to prefer opportunistic problem solving rather than full-fledged redirections and experiments. In Tom Anton's words, "patience, restraint, and accommodation" are the "latent norms of policymaking" in Sweden.[50]

Individual legislators trying to promote new policy ideas are also at a strategic disadvantage, given the institutional context of policy choice and change. Most of the policy-relevant information is provided by the agencies. Even if the agencies are formally independent, the cabinet has many ways of coordinating agency activities to suit cabinet preferences and policy objectives. By formulating the directives for Royal Commissions, the cabinet in fact determines what policy changes are to be considered. Since the agencies are in many ways closely related to the regulated interests (e.g., through interest representation on executive boards), the individual legislator can count on having a tough job getting those actors to share his view of what is practicable and feasible.

Furthermore, policy approaches have a tendency to become

powerful institutional constraints on future changes, once they are established. Agencies become strongly committed to existing programs, and see a strategic advantage in getting others to accept long-range extensions of these programs. A policymaker proposing changes that would reduce the agency's importance in the policy process—as would have resulted from changing to a system of environmental taxation—must count on meeting strong resistance from agencies and policymakers strongly committed to present policy approaches.

No wonder then that the majority of policymakers preferred to play it safe and opt for policy changes that they judged both progressive *and* practicable. No wonder then that the policymakers judged it more important to implement existing and only incrementally changed policies than to initiate new principles and drastic shifts in clean air policy. No wonder then that there was no strategic retreat on objectives, even when physical-environmental and socioeconomic conditions changed dramatically. Because of the almost programmatic suboptimization of the clean air policy, the tortoise's shell was hard enough to absorb the shocks.

For how long will this pattern of change and this style of changing prevail? Is it possible for clean air to remain a *valence* rather than a *position* issue in Swedish politics?[51] With energy policy casting its shadow over the tortoise's path, this seems impossible. As the parties are moving toward their final choices of future energy policies, they will have to reassess existing clean air policy and to take a stand on future air pollution control. They must decide whether to view the objective of clean air as a constraint on energy policy choice or to let energy policy objectives set the framework for clean air policy development. The choice also has great ramifications for Sweden's economy. Nuclear energy, which does not pollute the air, can be based on domestic uranium. This could ease the nation's balance-of-payment problem. A continued heavy dependence on oil, with subsequent sulfur pollution, will be a hard burden on Sweden's economy.

Energy policy is the most heavily stressed issue in Swedish politics in the latter part of the 1970s. As the voters become increasingly aware of the interdependence of the air pollution, energy, and economic policies, policymakers will undoubtedly be forced to take a more definite position on the future of Sweden's clean air policy.

In the end, the tortoise may thus find herself required to make more dramatic moves than the preceding race has taught her to make.

CHAPTER 9

The Hare or the Tortoise?

Polity and Politics

We began this investigation with one question and several underlying assumptions. We asked why policymakers in two highly industrialized nations—Sweden and the United States—have selected and pursued different policies to cope with common and technically similar problems of air pollution, and what the fates of these policies have been. One underlying assumption was that the problems are common and similar in the two nations. With regard to the physical-environmental and technological aspects of the air pollution problem, the assumption has been vindicated. It must be noted, however, that mobile source pollution was relatively more severe in the United States than in Sweden around 1970. We have found that policymakers in both countries have been aware of all technological innovations and alternatives to control air pollution. At first, we made no assumptions concerning the relative socioeconomic importance of the industries, activities, and behaviors generally associated with the generation of air pollution. Later, however, we were able to show that this socioeconomic importance was not significantly different in the two nations. This has not been used to suggest that the difference in size between the two economies and the difference in their relative dependence on the international global economy might not have been important to the clean air policymakers' choices. It has only been used to suggest that if socioeconomic considerations were uppermost in the minds of the policymakers at the time of policy choice, that choice certainly was not consistent with the socioeconomic conditions and consequences prevailing or foreseeable at the time.

Another crucial assumption guiding our investigation has been that it is not the physical-environmental, technological, or socioeconomic differences, but rather the political-institutional ones, that account for the differences in clean air policy choice and change.

These differences are indeed very significant. Sweden is a unitary state and a parliamentary democracy. The cabinet is formed by the majority in the Riksdag. Cabinet proposals are thus guaranteed a rather easy passage. The individual legislator is elected as one among several representatives from multimember constituencies. He is nominated by the party and represents the party more than the special interests of the constituency. Cabinet ministries and administrative agencies are formally separated from each other. In reality, however, there is close cooperation on most matters of policy interpretation and implementative guidelines. Special interests in Sweden are thoroughly organized. They are represented not only on the main vehicle for Swedish policy formulation, the Royal Commissions, but also on the boards of administrative agencies. Finally, it should be noted that the courts have no balancing power vis-à-vis the executive or legislative branches of government. Private interest groups thus cannot use the courts as a means for challenging governmental policy. In short, the center of gravity in the Swedish polity lies in the cabinet. It directs the Royal Commissions, it controls the legislature, and it has many levers to use against the bureaucracy.

The United States is a federal republic with a carefully balanced distribution of power between the executive, legislative, and judicial branches of government. The president is nationally elected, and his cabinet is not selected from the majority in Congress. Each individual legislator represents a single constituency. Adherence to the special interests of the constituents is as important as adherence to the party line, especially since there is no party line or party platform similar to the one facing a Swedish legislator. Congressional committees have an important policy-formulating and policy-evaluating means in the congressional hearing. The balance of power means that the legislators, contrary to the workings of the parliamentary democracy, will not become responsible for the implementation of selected policy alternatives. The balance of power further means that interest groups have several channels through which they can attempt to influence policy: the executive, the agencies, the legislature, and the courts. The courts can try the constitutionality of legislated policies and may influence policy implementation through interpretations and orders to administrative agencies. In short, there are several centers of gravity in the United States polity.

According to another of our underlying assumptions, systems do not themselves make policies, and polity differences do not in themselves make policy differences. But their importance lies in the fact that they provide different political contexts of choice. They provide different sets of political incentives and different sets of political constraints. In short, they provide for different types of clean air politics, different styles of clean air policy choice and change.

What links polity, politics, and policy in our analysis is the existence of rational policymakers capable of interpreting and assessing contexts, constraints, incentives, and choices in accordance with what they consider politically desirable and politically feasible.[1] Systems differences acquire an explanatory status in this comparative policy study only as they are found to play a decisive role as the actors' motivations and arguments for a particular policy action or policy content.

Proceeding along these lines, our investigation has identified two different politics of policymaking, two different styles of clean air policy choice and change. These two styles seem to correspond closely to the running styles of the two competitors in the ancient fable of the hare and the tortoise.

The United States style corresponded to that of the hare. Initially, policymaking proceeded very swiftly and dramatically, with many policymakers trying to take the lead by outbidding each other. The total period of policymaking was very short and compressed, and there was no need for compromises to build up a majority to secure the passage of legislation. Later on, policymaking proceeded at a much slower pace, and the total period of policymaking was very long and drawn out. Policymaking involved complicated efforts to build majorities around carefully designed compromises. These efforts included the organized pressure groups who had been very much left out in the open bidding for broad public support in the period preceding the initial policy choice.

Throughout the studied period, the Swedish style closely resembled that of the tortoise. There were no dramatic jumps but, rather, a slow and continuous movement. Not very many policymakers were involved, and they went to great lengths to find compromises that would make all four legs—the cabinet, the bureaucracy, the Riksdag, and the regulated interests—move in the same direction and with the same speed. The key words were compromise and consensus rather than competition, continuity rather than popularity.

From the perspective of policymakers as rational problem solvers, these two styles of policymaking are easily explained. The American policymaker is held personally responsible to his constituents. At the same time, he is relatively free of such constraints as party platforms and party commitments to certain policies. Furthermore, there is no chance that the legislator will become responsible for the actual implementation of policy. In an election year, and with a strong and seemingly unanimous public opinion pressing on a certain issue, the American policymaker thus faces strong incentives to make that issue a political one. He will find it strategically wise to join in the competition for visibility and leadership on popular issues by introducing bills, initiating committee hearings, etc. Since everybody competes for his own share of the popularity, policymaking will be short, compressed, and undaunted by compromise. This is exactly what happened in the process leading up to the 1970 Clean Air Act Amendments. In Senator Griffin's words: "When everybody is for clean air and against pollution, . . . it is difficult politically to vote for any amendment that would be characterized by the press as weakening the bill."

It follows that when "everybody" is no longer for clean air but is promoting issues seemingly contrary to clean air policies, then the policymaker will find it strategically wise to compete for policy leadership in that field. It may even be that if the popular standpoint is perceived to require an *attack* on the initial clean air policy, the policymaker will judge it strategically wise to do so. This is exactly what happened in the process leading up to the 1974 Energy Supply and Environment Coordination Act and the 1977 Clean Air Act Amendments. With the initial unanimity waning, those policymakers still heavily committed to clean air policy had to engage in time-consuming efforts to find the common denominators capable of attracting a majority of the policymakers. No longer was it perceived strategically beneficial to come out single-mindedly in favor of clean air. Leading policymakers consciously designed the politics of policymaking in such a way as to bring into consideration a broad set of relevant interests. The political strategy switched to accommodation rather than acclamation.

The Swedish policymaker has more to gain from going along with the party line than from trying to establish himself as the spokesman of a particular issue or constituency. Another constraint is the parliamentary system itself; to propose something which you and your majority fellows may not be able to implement is not very wise, since the cabinet is held responsible for bad administrative performance. At most, the Swedish policymaker will achieve the establishment of a Royal Commission. The whole idea of such commissions is to accommodate different interests, to find a compromise solution that satisfies all those involved. To paraphrase Tom Anton: "In Swedish politics, to accommodate is to survive."[2] The commission's report is filled with comprehensive reviews and data. Usually, the recommendations of the commissions are well anchored within the majority party or parties, the bureaucracy, and the interests concerned. In effect, it *is* the official policy, and individual policymakers will see little incentive in trying to break that policy agreement open. For where would they find the nonestablishment data and arguments needed to do so? At the time of the passage of the 1969 Environment Protection Act, there existed no environmental lobbying organization of the kind usually associated with modern environmentalism. Throughout the studied period, Swedish policymakers have been faced with a political context of choice that made it seem strategically beneficial to opt for consensus rather than conflict, problem solving rather than profile seeking.

To summarize, we have found that the political context is the most important of the contexts of choice in explaining the differences between Sweden and the United States in the field of clean air policy. We have found that the policymakers' rational assessment of the strategic incentives and constraints provided by the different political institutional contexts have given rise to two different dynamics of clean air policymaking.

In the United States clean air was quickly and easily made a political issue. Policy decisions were made in very swift fashion because policymakers wanted to join the popular environmental bandwagon. Later, there was less unanimity and more conflict. The speed of policymaking slowed down and became characterized by bargaining and deliberate efforts to build majorities around compromise formulas.

In Sweden clean air policymaking was for the most part a very long, drawn-out process. The style was less competitive, and the issue never became as divisive as in the United States. Swedish clean air politics has been marked by deliberate efforts to reach compromises acceptable to all interests concerned.

Politics and Policy

The importance of the political context of choice does not make itself felt only in the politics of clean air. We have found that the content of clean air policy choice and change is consistent with the policymakers' assessment of the strategic constraints and opportunities provided by their respective institutions of government and with the preferences revealed in their statements. In fact, we have found a close correspondence between the types of politics and the types of policy existing in the field of air pollution control.

In the United States, the first phase witnessed an obsession with radical goals and a seeming neglect for the availability of means to achieve these proud objectives. The competitive bandwagon atmosphere induced policymakers to engage in policy escalation and speculative augmentation beyond implementative capability. In the second more adversary and less unequivocal phase, policy content was marked by incrementalism, by efforts to adjust policy objectives gradually to available means and resources.

Swedish clean air policy has been marked by an obsession with means and resources rather than goals. Instead of going for what they *wanted* to do, the policymakers went for what they *could* do. Policy choice and change throughout has been marked by incrementalism, by a gradual adjustment of objectives to available means and resources.

This is especially true of the approaches to air pollution control as such and the technologies necessary to achieve cleaner air. However, this study also makes it clear that the policies of participation and power distribution are consistent with the actors' preferences and with a rational assessment of the institutional and cultural constraints and incentives operating within each of the political contexts of choice.

The 1970 United States approach to air pollution control can be summarized in three concepts: (1) statutory ambient air quality standards, (2) strict achievement deadlines, and (3) technology-forcing legislation. This was a tremendous escalation over earlier policies, but one deemed necessary by those who wanted to retain (or get) political leadership in a policy field characterized by strong public pressure for radical action. By relating clean air policy to public health rather than to overall environmental quality or actual existence of technological means for air pollution control, Edmund Muskie and others seemed to give a personal answer to all the Americans demanding clean air. By establishing numerical values for what should be regarded as clean air, and by setting up final dates for the achievement of these values, the policymakers tried to satisfy the strong public demand that needed a strong response.

The strong public demand also induced policymakers to take a tough line against the industrial and other socioeconomic interests affected by an escalated policy. At the height of his clean air rhetoric, Muskie told the auto manufacturers to make an automobile with which Americans can live by 1975-76. In their efforts to satisfy the strong wave of environmental opinion in the election year of 1970, American policymakers were ready to gamble. They assumed that technology would be forthcoming and that the American people would continue to accept the socioeconomic costs and burdens associated with air pollution control.

None of these assumptions proved accurate. After 1973-74, American policymakers engaged in what could best be described as policy deescalation. Health-related ambient standards were retained, but the deadlines were suspended time and again. This was an effort to adjust clean air policy to what policymakers now perceived to be socioeconomically feasible and technologically practicable. But above all, policy deescalation was a response to the drastic changes in the strategic context. The public wanted energy and employment more than environmental quality, and the business lobbyists regained their leverage with Washington policymakers. Leading policymakers, especially Paul Rogers, felt called upon to accommodate and carefully balance a wide range of conflicting interests to save what could be saved of existing clean air policy. The end result was a shift in policy content, from principles to practicability, from technology enforcement to technology adjustment. From the depths of despair, Muskie's chief aide lamented that General Motors, U.S. Steel, and others were right in thinking that they were above Congress and could force it to change.

However, some other important aspects of United States clean air policy never changed. The citizen suits introduced by Muskie in 1970 extended to all citizens the right to "trigger the enforcement mechanism." This extended participation was seen as necessary to guarantee good administrative performance, and has never been challenged. The introduction of this concept was a masterly move by Muskie in his efforts to satisfy public opinion and regain the initiative in the 1970 policy process. But it is also very much in line with fundamental principles of the American constitution and American political culture. Its durability in clean air policy is in large part explained by the fact that it is more an outgrowth of constitutional doctrine than of a passing environmental opinion.

The same is true for the distribution of powers in American clean air policy. The 1970 policy provided for federal leadership through the EPA but also put the states in a key position through the concept of state implementation plans. States' rights could not be violated, for both constitutional and strategic reasons. Developments after 1970 in such areas as transportation controls, nonattainment, and nondeterioration seemed to imply federal preemption. Much of the debate and decisions in 1976 and 1977 centered around efforts to retain or increase state powers and state involvement in clean air policy implementation.

The Swedish approach to air pollution control can be summarized in three concepts: (1) nonstatutory source emission guidelines, (2) diffuse achievement deadlines, and (3) adjustment to technological development. Unlike their American counterparts, Swedish clean air policymakers could formulate policy without involving public opinion in their calculi. The policy formulated between 1963 and 1969 preceded the growth of environmental opinion. Consequently, policymakers felt free to frame the issue in terms of conflicting uses of land rather than in terms of public health. They judged it strategically more important to find out what industry and bureaucracy actually could do, rather than to establish definite objectives expressing what would be necessary to protect public health. Therefore, there were at first no statutory ambient standards but only guidelines for maximum allowable emissions from individual sources. There were no specific achievement deadlines but rather language referring to economic feasibility and adjustment of policy objectives upward as technology continued to develop. For Swedish policymakers, it has remained much more important to adhere to existing standards than to promulgate new standards that could not be implemented. Swedish politicians know that strategic gains are not dependent on their ability to put new policies on the books but on their ability to make policies which actually achieve the intended results in the target area.³ Even the decision to establish more stringent auto emission standards than the rest of Europe fits the incremental adjustment pattern. Sweden applied the 1973 United States standards to 1976 car models. However, its largest business corporation, Volvo, already had developed a three-way catalyst meeting the 1975–76 U.S. standards!⁴

Typically enough, the health-related ambient air standards of 1976 were promulgated with little fanfare and without reference to what the health of the nation required. It was just another incremental step, taken as resources and knowledge became available and reliable.

The Swedish distribution of clean air policy powers and the provisions for public participation in policy implementation represent a peculiar balance of preference and strategic calculi. Swedish policymakers did not face such constraints as federalism or strong regional interests. Consequently, they felt free to centralize clean air policy powers at the national level in accordance with preferences for effectiveness and performance. It was only a last-minute strong Riksdag opinion stressing the need for due process that prevented an even greater centralization at the NEPB level. Leading policymakers also seemed to prefer performance to participation; in fact, good administrative performance would make participation unnecessary. Until this day, participation is limited to affected parties; citizenship has yet to become sufficient grounds for participation in the policy implementation process.

To summarize, we have found two distinct patterns of policy choice and change, which can be explained in terms of policymakers' rational assessment of the strategic constraints and opportunities perceived to be at work within their respective political systems. For, as we showed in chapters 6 through 8, physical-environmental, technological, and socioeconomic conditions did not differ so much as to render the different policy choices and changes consistent with a rational assessment of the constraints and opportunities offered by these contexts alone.

Thus, the United States hare initially set off with great speed. However, he soon became exhausted and began to lower his speed to fit his strength. His goal remained fixed, but he had to delay his arrival. The tortoise moved slowly, always adjusting her speed to fit her strength. She kept silent about the date of arrival, but she kept moving.

The ancient fable has a well-known moral about which way of running is most successful. As a last effort in this comparative endeavor, let us try to find out who is winning, or leading, in the race for clean air.

Policy and Impacts

It would of course be easy to say that the Americans obviously did not achieve ambient air quality standards at the original date. It would also be easy to say that they certainly did not achieve the goal with respect to auto emissions. But such an analysis would beg the question; it would give us no real comparison with Sweden, since that country has not established any clear-cut ambient air quality standards or specified any deadlines for achieving them.

The assumptions made in chapter 2 about the preconditions for successful policy implementation are thus very difficult to test. The policy objectives can be determined with certainty only for the United States but not for Sweden. Coupled to the Swedish search for feasibility and practicability was an almost programmatic unwillingness to get associated with precise air quality levels or fixed achievement deadlines.

Not only are the American objectives and deadlines for achievement spelled out more clearly, they are also associated with an almost totally different set of policy concepts. The United States approach is geared to achieving specified levels of ambient air quality, that is, specified concentrations of certain pollutants in ambient air. The control of individual sources is closely connected to the ambient air quality levels. In Sweden, the emphasis is on controlling the emissions from each individual source. The guidelines for source emission control have not been explicitly coupled to an ambient air quality approach. As a result, one is confronted with two types of air quality data. American reports concentrate on ambient air quality data and are based on a nationwide monitoring system. Swedish data are based on reports from individual sources, and a nationwide monitoring system will not become a reality until the early 1980s.

So the analysis of clean air policy impacts is faced with several problems inherent in impact analysis in general, as well as with problems inherent in comparative analysis. The impact analyst must somehow find his own standard of judgment. But the choice of such a standard is constrained in several ways: (1) It must be a standard that is universally applicable to all cases. (2) Since this is a comparison of national policies, the standard must be applicable at the national level. (3) Furthermore, the standard of judgment must be associated with some major problem or objective in the policy field. Since this is an analysis of politics, policy, and impact, we must somehow find a standard of judgment which implies the possibility of a political feedback from the original policy impact.

The standards of judgment used here are average citizens and their exposure to air pollution emissions. Citizen exposure is a universally applicable concept and can be used for national comparisons. The health of the citizen has been the most important objective in the United States policy, and is more and more coming to the fore in Swedish clean air policy. The impact on citizens has the potential of a future political feedback that may lead to changes in existing policy.

But the problem of impact analysis does not end there. Even if we find some interesting patterns and differences, how can we be sure that the policy and nothing else caused these differing impacts? As we will soon find out, this *ceteris paribus* problem is present also in our impact analysis.[5]

How then is the impact analysis made? First, let us have a look at the available data. Both countries have presented nationwide emission estimates. The United States data are more complete than those from Sweden, since they also include a miscellaneous source category. This category will be excluded here, since we want to concentrate on manmade air pollution. Estimates concerning stationary source emissions seem to be based on the same types of information: " . . . published data on fuel use and industrial production, other . . . data such as air pollution emission factors, and available information on the extent of air pollutions employed."[6] As for Sweden, the data are also based on actual measurements at industrial plants.

Swedish estimates of total auto emissions are based on assumptions concerning the average driving pattern, the results of exhaust emission tests, and knowledge about the existence and distribution of different emission control technologies among Swedish vehicles. These data are used to compute the average emissions per vehicle kilometer. The number of vehicles is easily found in official statistics. With these data, and with figures for average vehicle travel based on survey and statistical data, the NEPB makes its estimates of total auto emissions in Sweden.[7] The EPA report on emission trends of December 1977 gives no clue as to what methods have been used to compute total auto emissions. To get comparable figures, I made use of the figures presented in the 1977 *Statistical Abstracts of the United States* concerning average emissions per vehicle mile, average vehicle travel, and the number of vehicles registered each year.[8] With these data, and with the population figures for each country, I computed a series of indices for citizen exposure to air pollution emissions. In this way, I tried to make the data for the two countries as comparable as possible. The measuring rod is the individual citizen's pollution load. By constructing indices, the hare and the tortoise are placed at the same starting point. Let us follow the race!

The index for total air pollution emissions does not reveal any remarkable differences between the two countries (see fig. 12). The average Swede's pollution load fell off by about 8 percent between 1970 and 1976, while the average American's pollution load decreased by about 10.5 percent during the same period.

Just as expected, the Swedish pattern shows no dramatic upward or downward movements. There is a steady, albeit moderate yearly decrease. From earlier chapters we might have gotten the impression that the United States pattern would initially be a dramatic decrease followed by a more stable development. Our figure reveals no such

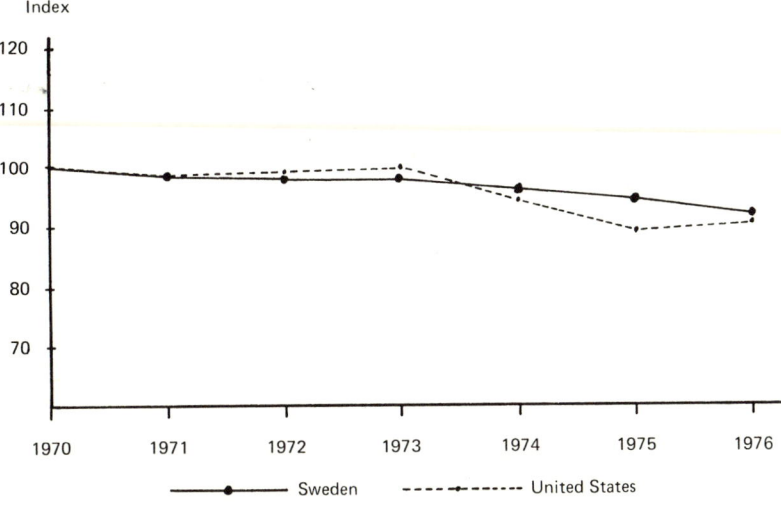

Fig. 12. Index for total air pollution emissions in Sweden and the United States, 1970-76, in kg/inhabitant (Sources: Sweden—*Statistisk Årsbok* 1977, pp. 30, 193; Bilismen i Sverige 1977, pp. 22 ff.; NEPB report to the OECD, December 1977. United States—*U.S. Statistical Abstracts* 1977, pp. 28, 205 f., 634; EPA, *National Air Quality and Emissions Trends Report* 1976, December 1977, pp. 5-1 ff.)

trend. Instead, the pollution load increases slightly up to 1973, decreases between 1973 and 1975, and then goes upward.

Still, this could be seen as an effect of changes in the policy. The 1970 Clean Air Act Amendments began to be fully implemented by 1973. Then the state implementation plans took effect, and more stringent auto emission standards were imposed. But such an interpretation leaves out some important facts about the realities of policy implementation. That the implementation plans took effect in 1973 did not mean that all the sources immediately began to emit less air pollutants than before. The states had to engage in negotiations with industries to establish compliance schedules. The firms had to install the pollution control equipment necessary to follow the schedules. Furthermore, the 1973 auto emission standards applied only to new car models, amounting to about 10 percent of the total car fleet each year. These factors imply that we should be careful not to jump to any affirmative conclusions concerning the relationship between policy and impact.

Figure 13 shows that there are different emissions trends despite the seemingly uniform overall pattern. It shows further that the differences are in line with what we could expect from the differences in policy

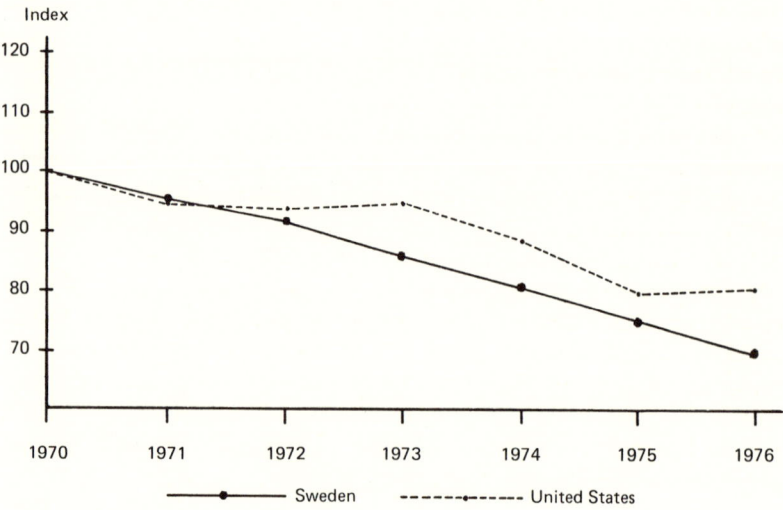

Fig. 13. Index for air pollution emissions from other sources than highway vehicles in Sweden and the United States, 1970–1976, in kg/inhabitant (Sources: see fig. 12)

emphasis. The Swedish policy was primarily focused on controlling emissions from individual sources. State subsidies were given to existing plants for installing pollution control equipment. As a result, emissions of particulates have been cut in half, and sulfur dioxide emissions have decreased by more than 20 percent.[9]

The United States policy has not involved economic subsidies as a means to lure reluctant industries to install scrubbers and other pollution control equipment. One is left with the impression that industrial cooperation to fight air pollution has been more difficult to obtain than in Sweden. Yet, the emissions of particulates have been cut by 40 percent. On the other hand, total emissions of sulfur dioxide have not been reduced by more than 8 percent.[10]

Can we be sure that these reductions are direct effects of the clean air policies? Since air pollution from stationary sources is directly related to industrial and energy production, it may well be that variations in production activities cause variation in total emissions. In the Swedish case, there seem to have been no such variations during the period. The index for industrial production rose every year from 1970 to 1976, probably as a result of the government's deliberate efforts to bridge Sweden's economy over the international recession following the oil embargo.[11] In the United States, the industrial production index levelled off in 1974 and was way down in 1975. Even the EPA concluded that "emissions in 1975 were generally lower than those in 1976 because of economic conditions in 1975 that reduced industrial production. Increased emissions from 1975 to 1976 reflect the effects of economic recovery."[12]

We may thus conclude that both the Swedish and the American policies have had an impact on stationary source emissions. The Swedish policy seems to have been somewhat more effective, since some of the American reduction is clearly related to the decrease in industrial activity arising from the economic recession of the mid-1970s.

The United States policy of 1970 was especially tough on auto emissions, while the Swedish policy has been one of adjusting standards to technological development. United States policy measures have been introduced three to four years later in Sweden. Quite naturally, one would then assume a far greater reduction in auto emissions in the United States than in Sweden. Figure 14 evidently corroborates this assumption. The Swedish regulations of 1969, 1971, and 1976 were not enough to offset the increase in the number of vehicles. Average emissions per vehicle kilometer decreased by 8 percent, while the number of vehicles increased by 25 percent. Consequently, the average Swede has become increasingly exposed to emissions from highway vehicles.[13]

In the United States, average emissions per vehicle mile decreased by 18 percent between 1970 and 1976. The decrease in total auto emissions, and thus in citizen exposure, seems to have been especially marked in 1973–74, after the imposition of more stringent auto emission

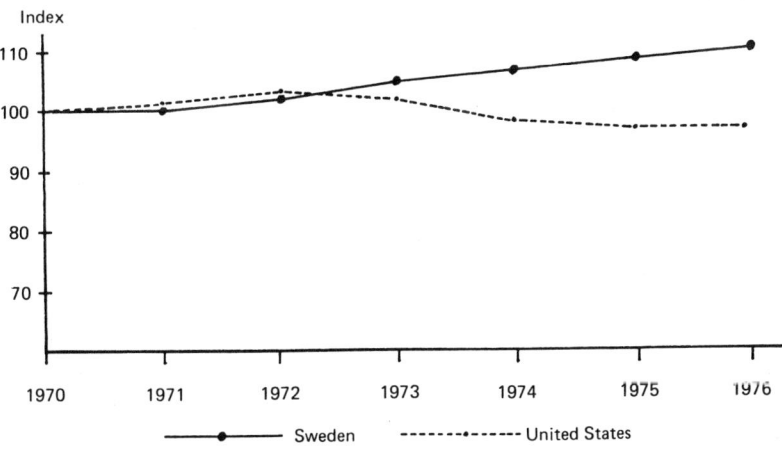

Fig. 14. Index for air pollution emissions from highway vehicles in Sweden and the United States, 1970–1976, in kg/inhabitant (Sources: see fig. 12)

standards. On the face of this, one might be tempted to conclude that United States policy has had a greater impact than its Swedish counterpart. But the 1973–74 dip itself should make us extremely cautious. Other factors might well be at work here. First of all, average emissions per vehicle mile do not show the same dip for 1973–74. Secondly, the increase in the number of vehicles during 1973–74 was as large as the increase during 1972–73. Given the fact that the percentage increase in the number of vehicles was larger than the percentage decrease in average emissions per vehicle mile, and given further the fact that the new 1973–74 vehicles represented less than one fifth of the total vehicle population, one must conclude that the sudden decrease in the pollution load from auto emissions did not come just from the new auto emission standards.

The key variable seems to be the average number of vehicle miles traveled each year. That number increased up to 1972, then fell off somewhat in 1973, but fell way down in 1974. By 1976 it had yet to reach the level of 1970. Evidently the oil crisis, the gasoline shortage, and the subsequently skyrocketing gasoline prices made Americans think twice before using the car as much as they had done before. This shortened travel seems to account for most of the decrease in auto pollution after 1973. Had the average number of vehicle miles been the same in 1976 as in 1970, the average pollution load on every citizen would also have been the same.[14]

This, however, gives us reason to state that the American policy has been more effective than the Swedish one. For it means that the United States policy has been able to keep the average citizen's exposure to auto emissions from increasing despite the increase in the number of vehicles. It further means that the United States policy would have had this impact even if there had been no decrease in average vehicle travel. The Swedish policy has not been that successful. There has been no increase or decrease in average vehicle travel. As we have seen, however, the imposition of new and gradually more stringent auto emission standards has not sufficed to offset the increased pollution load following from an ever-larger number of motor vehicles.

Our analysis makes it clear that it is very difficult to pick the winner at this stage in the race. In fact, it would seem as if we have two different races going on at the same time, with two different leaders. The tortoise seems to be doing a better job in controlling emissions from stationary sources, while the hare is in the lead when it comes to controlling mobile sources.[15] At this point, the overall result seems very much like a dead heat. Neither of the two countries seems to have been remarkably more successful than the other in relieving its citizens from the blight of air pollution.

Conclusion

The fact remains: both countries have experienced some success in fighting air pollution. To the reader, this may seem like a surprise. We have stressed that Sweden has always made sure that selected policy objectives are matched by implementative resources. We have shown how regulated interests have been deliberately incorporated into the implementation process. Somehow, this seemed to indicate that the Swedish clean air policy would be more successfully implemented and would have a more noticeable impact than its American counterpart. Not only did the Americans escalate policy beyond implementative capabilities, but they also changed the policy ambitions downward as the implementative problems began to make themselves clearly felt. This may have conveyed the impression that the American policy never really got out of the blocks and thus never had any impact on air pollution levels.

However, this is only half the picture. As we pointed out in chapter 2, policy implementation is a different game than policymaking. That is especially true for the United States, where the implementing agency is faced with a context of choice that differs substantially from the one facing American policymakers in Congress. The 1970 Clean Air Act Amendments gave American citizens the right to sue the agencies to trigger policy enforcement. Thus the courts were given a pivotal position in the game.

The courts are supposed to let only substantive calculations and the preferences spelled out in existing legislation form the basis for decisions. We must keep in mind that what guided court opinions throughout most of the period was the 1970 act with its far-reaching intentions and clear-cut criteria. As John Quarles makes vividly clear in his account of the EPA's first years, the agency was constantly haunted by court orders to make implementation more consistent with legal interpretations of legislative intent. Such court orders came even as the EPA was in the midst of painstaking strategic efforts to have the policy moderated in accordance with the contextual changes following the 1973 oil embargo.

We may conclude that the context of choice in United States policy implementation left policy administrators with little choice but to try to implement the intentions of the 1970 amendments. The changes made in 1973–75 concerned deadlines for achievement, but not the principles and intentions of the policy. No wonder then that we find a more effective implementation than the proposition stated in chapter 2 made us believe.

How different the context of policy implementation in Sweden!

Undisturbed by citizen suits and court orders, the Swedish administrators could engage in negotiations with polluters to find an acceptable formula for policy implementation. Furthermore, these negotiations were not constrained by any legally prescribed standards or compliance deadlines. This should not be interpreted as implying that there have been any deviations from legislative intent. However, one is left with the impression that this consensual and cozy context provides less incentive for vigorous enforcement than does the adversary American context.

If we return finally to the propositions stated in chapter 2, we find that only their first half is corroborated by this study. The American institutional context of choice did provide policymakers with incentives to make the clean air issue a political issue. It was easily brought to the political agenda and became subject to political conflict and controversy. The Swedish institutional context of choice did not provide policymakers with much incentive to politicize the clean air issue. When it was brought to the agenda, it never became subject to much conflict and controversy.

We must conclude that the second part of these propositions is not corroborated by our study. Contrary to our assumptions, the United States clean air policy has been subject to continued implementation, thanks in great part to the special context of constraints and incentives provided by the 1970 legislation itself. As the courts continued to interpret the legislative intent of the 1970 act, they forced administrators to implementative action. In Sweden the close cooperation between the agency and the regulated interests may have led to a less vigorous and less effective implementation than would have been the case had the NEPB been subject to citizen suits and court orders.

Thus, the realities of modern politics and policy are not wholly consistent with the teachings of the ancient fable. In political analysis, fables may be suggestive, but they are seldom, if ever, conclusive.

Notes

Chapter 1

1. For a detailed account of the Swedish context and process of choice, see Lennart J. Lundqvist, *Miljövårdsförvaltning och politisk struktur* (Lund: PRISMA/Verdandidebatt, 1971).
2. The American process is described at length in Walter A. Rosenbaum, *The Politics of Environmental Concern* (New York: Praeger Publishers, 1973), chap. 5. The best account is Charles O. Jones, *Clean Air: The Policies and Politics of Pollution Control* (Pittsburgh: University of Pittsburgh Press, 1975).
3. Environmental Protection Agency, *A Progress Report* (Washington, D.C.: EPA, 1972), p. 1; cf. Environmental Protection Agency, *The Challenge of the Environment: A Primer to EPA's Statutory Authority* (Washington, D.C.: EPA, 1972), p. 6.
4. Environmental Protection Agency, *Progress in the Prevention and Control of Air Pollution in 1973. Report to Congress* (Washington, D.C.: EPA, 1974), p. 24.
5. *Congressional Quarterly Almanac,* 1970, p. 474. Cf. EPA, *A Progress Report,* p. 1.
6. Environmental Protection Agency, *Progress in Prevention and Control,* p. 50.
7. *Congressional Quarterly Almanac,* 1970, p. 472.
8. Environmental Protection Agency, *Progress in Prevention and Control,* p. 80.
9. Göran Persson, "Synpunkter på luftvårdslagstiftningens tillämpning i Sverige," in *Luftvård 73: Konferens om nordisk luftvårdspolicy, Helsingfors 21-22 November 1973* (Helsingfors: NORDFORSKS miljövårdssekretariat, 1974:6), p. 51. Mr. Persson was then head of the air quality department within the NEPB.
10. *Swedish Code of Statutes* 1969:387, §§1, 2, 4–6, 8, 9, 18. The act, with commentaries in English, is found in Royal Ministry for Foreign Affairs et al., *Environment protection act. Marine dumping prohibition act. With commentaries. Information to the United Nations conference on the human environment* (Stockholm: ALLF, 1972).
11. NEPB, *Naturvårdsverkets årsbok 1973* (Stockholm: ALLF, 1974), pp. 77 ff. For information on the numerical values of the guidelines, see NEPB, *Riktlinjer för luftvård* (Stockholm: NEPB, 1973:8).
12. *Swedish Code of Statutes* 1969:388, §§1–8.

13. *Swedish Code of Statutes* 1969:397, §§10, 18, 22, 24.
14. *Swedish Code of Statutes* 1972:782, §136a.
15. A preliminary evaluation of the grants program is found in NEPB, *Naturvårdsverkets årsbok 1973,* pp. 82 ff.
16. Environmental Protection Agency, *Progress in Prevention and Control,* pp. 25 ff.
17. Environmental Protection Agency, *A Progress Report,* p. 3; NEPB, *Naturvårdsverkets årsbok 1973,* pp. 77 f.
18. Environmental Protection Agency, *Progress in Prevention and Control,* pp. 6 ff.; cf. Council on Environmental Quality, *Environmental Quality: The Fourth Annual Report of the Council on Environmental Quality* (Washington, D.C.: Govt. Printing Office, 1973), pp. 163 ff.
19. *Swedish Code of Statutes,* 1972:596; NEPB, *Naturvårdsverkets årsbok 1972* (Stockholm: ALLF, 1973), p. 82.
20. NEPB, *Naturvårdsverkets årsbok 1973,* p. 14; Environmental Protection Agency, *Progress in Prevention and Control,* p. 76 f.; NEPB, *Naturvårdsverkets årsbok 1972,* pp. 84 f.
21. NEPB, *Miljövård i Sverige: lagstiftning, administration, forskning, anslag* (Stockholm: ALLF, 1972), passim; cf. Royal Ministry for Foreign Affairs, et al., *Sweden's national report to the United Nations on the human environment* (Stockholm: Norstedt & Söner, 1971), pp. 58 ff.
22. Council on Environmental Quality, *Environmental Quality. The Second Annual Report of the Council on Environmental Quality* (Washington, D.C.: Govt. Printing Office, 1971), pp. 6 ff.
23. Ibid., pp. 4 ff.; Environmental Protection Agency, *A Progress Report,* pp. xiii ff.
24. Environmental Protection Agency, *A Progress Report,* pp. 5, 97.
25. NEPB, *Naturvårdsverkets årsbok 1973,* p. 117. Cf. ibid., pp. 31 ff.
26. Environmental Protection Agency, *Progress in Prevention and Control,* pp. 51 ff.
27. Cabinet Proposal 1974:46 med riktlinjer för utvecklingen av miljövårdens informationssystem.
28. Environmental Protection Agency, *Progress in Prevention and Control,* pp. 6 ff.
29. Cf. the discussion found in NEPB, *Miljövård i Sverige: lagstiftning, administration, forskning, anslag,* p. 17; cf. NEPB, *Naturvårdsverkets årsbok 1972,* pp. 82 ff. Emission regulations are primarily handled by the National Traffic Safety Board (NTSB).
30. Environmental Protection Agency, *Progress in Prevention and Control,* pp. 32, 40 ff.; idem., *The Challenge of the Environment: A Primer to EPA's Statutory Authority,* p. 10.
31. Ministry for Foreign Affairs et al., *Environment protection act. Marine dumping prohibition act. With commentaries,* p. 35.
32. Ibid., pp. 63 ff.
33. Since 1969, only a handful of prohibitions have been issued. About twenty cases of compensation have been brought before the real estate courts, and several have been withdrawn and settled outside of the courts; cf. OECD

Environment Directorate, "Environmental Policy of Sweden in Relation to the 'Guiding Principles': The Environment Protection Act and Related Legislation" (Paris: OECD, Environment Directorate, September 1973), pp. 8 f.
34. Environmental Protection Agency, *A Progress Report*, p. 89; see also idem., *The Challenge of the Environment*, p. 10, and *Progress in Prevention and Control*, pp. 38 f.
35. Environmental Protection Agency, *The First Two Years—A Review of EPA's Enforcement Program* (Springfield, Va.: NTIS, February 1973, pp. 1, 243 ff.; idem., *Progress in Prevention and Control*, p. 37.
36. R. E. Ayres and J. F. Miller, *Citizen Suits under the Clean Air Act Amendments of 1970* (Washington, D.C.: EPA, n.d.), pp. 2 f.; cf. Council on Environmental Quality, *Environmental Quality: The Fourth Annual Report of the Council on Environmental Quality*, p. 395.
37. Ministry for Foreign Affairs et al., *Environment protection act. Marine dumping prohibition act. With Commentaries*, pp. 66, 74 ff.
38. *Congressional Quarterly Almanac*, 1970, p. 483 f.; EPA, *A Progress Report*, p. 2; cf. footnote 37.
39. NEPB, *Naturvårdsverkets årsbok 1973*, p. 77.
40. Environmental Protection Agency, *Progress in Prevention and Control*, pp. 57 ff.
41. Cf. NEPB, "Miljövårdseffekten av statsbidragen till miljövårdande åtgärder" (NEPB PM 488, July 29, 1974). PM 488 states that the volumes of contained emissions are estimated "partly on expected results, partly on inspection protocols." Thus, a large part of the volumes is what the firm *expects* to get out of its investment in protective measures.
42. Environmental Protection Agency, *The First Two Years—A Review of EPA's Enforcement Program*, p. 2.

Chapter 2
1. As I understand them, the following two books are representative of this type of causal policy analysis: Thomas R. Dye, *Policy Analysis. What Governments Do, Why They Do It, and What Difference It Makes* (University, Ala.: The University of Alabama Press, 1976), and Richard I. Hofferbert, *The Study of Public Policy* (Indianapolis: The Bobbs-Merrill Co., 1974).
2. Cf. the perspective taken in Charles W. Anderson, "System and Strategy in Comparative Policy Analysis: A Plea for Contextual and Experiential Knowledge," in *Perspectives on Public Policy-Making*, ed. William Gwyn and George C. Edwards III, Tulane Studies in Political Sciences, vol. 15 (New Orleans, 1975), pp. 219 ff.
3. Dean Mann, "Environmental Policy," *Policy Studies Journal* 1 (1972): 17.
4. Cf. the definitions offered in James E. Anderson, *Public Policy-Making* (New York: Praeger Publishers, 1975), pp. 2 ff.
5. Charles W. Anderson, "System and Strategy," p. 227.
6. Ibid., p. 234.
7. I owe this point to the discussion presented by my colleague, Sverker

Gustavsson, in "Types of Policy, Types of Politics" (unpublished manuscript), where he points to the different implications of the Lowian and Wilsonian schemes for classification of policy content.
8. Charles W. Anderson, "Comparative Policy Analysis: The Design of Measures," *Comparative Politics* 4 (1971): 129 f.
9. Cf. Charles W. Anderson, "System and Strategy," p. 226.
10. Ibid., p. 231.
11. Cf. Evert Vedung, *Det rationella politiska samtalet. Hur politiska budskap tolkas, ordnas och prövas* (Stockholm: Aldus/Bonniers, 1977), pp. 19 ff. Evidently, the strategic calculus can be seen as containing two elements, one descriptive, the other normative. The descriptive part consists of actor A's assessment of what the other actors will do if A proposes a certain alternative. The normative part contains actor A's assessment of whether or not he wishes that strategic consequence to occur. I am indebted to my colleague, Sverker Gustavsson, for this point. For all practical purposes, the distinction will not be specifically pursued here.
12. With respect to the strategic considerations of policymakers, one cannot of course expect to find much in the way of explicit references in the key documents. Thus, I will rely on different accounts and different sources—other books, articles and scientific works, media accounts, etc.—to find out what informal-relational strategic considerations might have been entertained by key policymakers at the time of choice.
13. Cf. Charles W. Anderson, "System and Strategy," pp. 233 f.
14. I am aware of the close relationship between formal political institutions and dominant features in a nation's political culture. For a discussion of political culture as a constraint on the adoption of policy alternatives, see Anthony King, "Ideas, Institutions, and the Policies of Government: A Comparative Analysis," *British Journal of Political Science* 3 (1973): 418 ff.
15. On the conditions for periodic policy swings in the U.S. balance-of-power system, see Arnold Heidenheimer et al., *Comparative Public Policy. The Politics of Social Choice in Europe and America* (New York: St. Martin's Press, 1975), p. 262.
16. Cf. John G. Grumm, "The Analysis of Policy Impact," in *The Handbook of Political Science*, vol. 6, ed. Fred I. Greenstein and Nelson W. Polsby (Reading, Mass.: Addison-Wesley Publishing Co., 1975), pp. 448 f.
17. Aaron Wildavsky, "The Strategic Retreat on Objectives," *Policy Analysis* 2 (1976): 520 ff.
18. Cf. the argument put forth by George C. Edwards III, "Congressional Responsiveness to Public Opinion: A Policy Perspective," *Policy Studies* 5 (1977): 488.
19. See, for example, Jones, *Clean Air*, chap. 7–8; Donald R. Kelley et al., *The Economic Superpowers and the Environment: The United States, The Soviet Union, and Japan* (San Francisco: W. H. Freeman & Co., 1976), pp. 200 ff.; Cynthia H. Enloe, *The Politics of Pollution in a Comparative Perspective* (New York: David McKay Co., 1975), pp. 324 ff.; William Solesbury, "Issues and Innovations in Environmental Policy in Britain, West Germany and California," *Policy Analysis* 2 (1976): 36 f. The results

of these and other studies are discussed and summarized in hypothetical form in my "The Comparative Study of Environmental Politics: From Garbage to Gold?" *International Journal of Environmental Studies* 11 (1978): 89-97.
20. Neil J. Smelser, "The Methodology of Comparative Analysis," in *Comparative Research Methods,* ed. Donald P. Warwick and Samuel Osherson (Englewood Cliffs, N.J.: Prentice-Hall, 1973), pp. 72 ff.
21. Donald P. Warwick and Samuel Osherson, "Comparative Analysis in the Social Sciences," in Warwick and Osherson, *Comparative Research,* pp. 22 f.; cf. Smelser, "Methodology," p. 74, where he argues that if the dimensions chosen are in principle universal, and if the principles of operational definition are worked out in accord "with the variety of social goals and meanings to which these dimensions are related," the problem of equivalence and comparability are, in fact, overcome.
22. For further discussions of the characteristics and canons of comparative method, see Evert Vedung, "The Comparative Method and Its Neighbours," in *Power and Political Theory: Some European Perspectives,* ed. Brian Barry (London: Wiley, 1976), pp. 201 ff.; Arend Lijphart, "The Comparable Cases Strategy in Comparative Research," *Comparative Political Studies* 5 (1975): 158 ff.

Chapter 3

1. SOU 1966:65, *Luftförorening, buller och andra immissioner* (Stockholm: Ministry of Justice, 1966), pp. 147 ff.
2. The following account is built on cabinet proposal 1969:28, pp. 178-220, passim; quot., p. 180.
3. Also in spring 1969, the cabinet proposed and the Riksdag authorized a five-year program of subsidies to environmental and pollution-control investments in Swedish industry. According to the program, subsidies would cover 25 percent of investment costs, and go to investments in old, existing plants and factories only. Cabinet proposal 1969:1:11, K 10.
4. Motions I:868 and II:482 (Liberal, on the necessity for a strong environmental policy program); I:946 and II:1086 (Liberal, on the necessity for more explicit guidelines for administrative decision making); I:393 and II:455 (Conservative, on public health and the effects of air pollution); I:947 and II:1083 (Center party, on the necessity for more explicit guidelines). Committee report, LU³ 1969:37, pp. 143-172, esp. pp. 162 ff. (minority views). *Riksdag Record,* AK 1969 28:81 ff. (Mrs. Anér, Liberal); 28:78, 86 f. (Mrs. Sundberg, Mr. Wachtmeister, Conservatives).
5. See SOU 1966:65, pp. 155 ff. See also my *Miljövårdsförvaltning och politisk struktur* (Lund: PRISMA/Verdandidebatt, 1971), pp. 192 ff.
6. See ibid., pp. 194 ff., for a description and analysis of the composition and functioning of the joint groups of experts.
7. SOU 1966:65, op. cit., pp. 226 ff.; Göran Persson (Head of Air Quality Department, NEPB), "Different approaches to air pollution control," paper for WHO interregional symposium on air quality criteria and guidelines, 5-9.10, 1970.
8. NEPB, *Förslag till riktlinjer för emissionsbegränsande åtgärder vid*

202 Notes

luftförorendande anläggningar and *PM med kommentarer till riktvärdesförslag augusti 1969* (Stockholm: NEPB, 1969), pp. 10, 15, 22.
9. Cf. Lundqvist, *Miljövårdsförvaltning och politisk struktur,* p. 209 (NEPB comments); p. 210 (comments from the pulp and paper industry representative).
10. Ibid., p. 210 f. Representatives of environmental hygiene wanted the same relations with the NEPB as those enjoyed by Swedish industries.
11. Royal Ministry of Foreign Affairs, et al., *Air pollution across national boundaries. The impact on the environment of sulfur in air and precipitation. Sweden's case study for the United Nations conference on the human environment* (Stockholm: Norstedt & Söner, 1971).
12. The NEPB's proposal and arguments are found in cabinet proposal 1968:122, pp. 5 ff. The time schedule was as follows:

Date	Maximum permissible SO_2-emission (gm/kg oil)	Equivalent sulfur content in oil (% weight)
July 1, 1969	50	2.5
July 1, 1970	40	2.0
July 1, 1972	30	1.5[1]
July 1, 1974	20	1.0[1]

[1] Desulfurization necessary

13. Industrial comments in cabinet proposal 1968:122, pp. 9 ff.
14. Minister's comments, ibid., pp. 12 ff.
15. Motions I:972 and II:1234 (Liberal); I:28 and II:37, and II:1227 (Leftist Communist); I:378 and II:470, I:967 and II:1226 (Conservative). The cabinet proposal was commented on by the Third Legislative Committee in the Riksdag, LU[3] report 1968:64. The Conservative motions on environmentally geared energy taxation were discussed and reported by the Ways and Means Committee, BeU report 1968:70, pp. 16 ff. The latter committee stated as a principle that a "new special environmental tax system is not in accordance with the establishment of such a general value added tax that business has wanted and the Riksdag enacted," p. 18. See also the *Riksdag Record,* AK 1968 37:139, and AK 1968 45:83 ff.
16. The Minister's memorandum is quoted in LU[3] report 1966:25, p. 2.
17. Motions 1966 I:123 and II:167 (Liberal); Committee report LU[3] 1966:25; *Riksdag Record,* AK 1966 17:49 ff. (Mr. Wiklund, Liberal), 17:51 (Mr. Lewin, Social Democrat); *Riksdag Record,* AK 1967 24:14 ff. (Olof Palme, Minister of Transportation), 24:17 (Mr. Wiklund, Liberal).
18. The specialist group had reported on this matter in March 1967 and proposed the legislative change. See cabinet proposal 1967:166, p. 32; Minister's arguments, see ibid., p. 49.
19. The main features of the specialist group's report, *Avgaser från bensindrivna bilar, utredning med förslag till åtgärder* (Stockholm: Ministry of Transportation, 1968, mimeo K 1968:2), are found in cabinet proposal 1968:160, pp. 11–24. According to the group, the number of cars

in Sweden was 2,030,000 in 1966. Average gasoline consumption per car was 1,600 liters. The group estimated the *total auto exhaust emissions for 1968* (figures given are in tons per year): Carbon monoxide (CO), 850,000; Hydrocarbons (HC), 140,000; Nitrogen oxides (NO_x), 40,000; Lead compounds, 2,500.

The group proposed a three-step program of emission performance standards:

Alternative	Applicable model year	Maximum emissions (gm/km)		Estimated % reduction of emissions	
		CO	HC	CO	HC
1	1971	45	2.2	40	40
2	1973	30	1.8	60	50
3	1975	23	1.5	70	60

Alternative 1 represented the minimum requirement applicable to 1971 models. It would cost 150 million Swedish crowns for 1971 and 1972 and require no major technological development. Alternative 2 corresponded to U.S. 1968 CO standards, but was 25 percent less stringent than U.S. 1968 HC standards. It would cost 260 million crowns and require major technological changes in many European car engines. Alternative 3 required 25 percent better CO performance than U.S. 1970 standards, and was equivalent to these U.S. standards for HC. It would cost 580 million crowns for 1975 and 1976 and require "such essential changes in most present engine types that a long lead time for development and testing will be necessary."

20. Car dealers' comments are found in cabinet proposal 1968:160, pp. 27 ff. Minister's argument, ibid., pp. 32 ff. Committee report, LU^3 1968:70; *Riksdag Record,* AK 1968 43:24.
21. See committee report LU^2 1968:55, pp. 2 ff. Existing statutes maximized lead content to 0.8 grams TAL per liter gasoline. The mean lead content for gasoline marketed in 1968 was 0.67 grams per liter. The specialist group's regulatory scheme was envisaged as follows:

Date	Maximum allowable lead content (gm/liter)
January 1, 1970	0.7
January 1, 1971	0.6
January 1, 1973	0.5
January 1, 1975	0.4

22. Motions 1968 I:45 and II:70 (Social Democrat). Committee report, LU^2 1968:55, pp. 5 f., *Riksdag Record,* AK 1968 37:98 f. The Poisons and Pesticides Board's decision is commented on in NEPB, *Naturvårdsverkets årsbok 1969,* p. 77.

204 *Notes*

23. The president's message, together with all other central policy documents, is reprinted in U.S. Senate Committee on Public Works, *A Legislative History of the Clean Air Amendments of 1970,* with a section-by-section index, prepared by the Environmental Policy Division of the Congressional Research Service of the Library of Congress for the Committee on Public Works, U.S. Senate, 93d Congress, 2d Session (Washington, D.C.: Govt. Printing Office, 1974), vols. 1 and 2. These volumes will be referred to henceforth as *A Legislative History* (1) or (2). See pp. 1502 ff. in (2) for the president's message on air pollution. The administration-sponsored bill S. 3466 is found on pp. 1474 ff.; the specific character of the new measures is discussed on pp. 1495 ff., "Summary of Clean Air Act Amendments of 1970." For provisions concerning technological feasibility, see S. 3466, pp. 1485 f. and 1489 f. The existing and recommended auto emission standards were as follows (ibid., p. 1325):

Exhaust	Model year (gm/vehicle mile)			
	1968	1970	1973	1975
HC	3.3	2.2	2.2	0.5
CO	34.0	23.0	23.0	11.0
NO_x			3.0	0.9
Particulates				0.1

24. *A Legislative History* (2), pp. 891, 895 ff., report on H.R. 17255.
25. Ibid., pp. 891 ff. The committee strengthened the administration's proposals concerning national ambient air quality standards by requiring the Secretary of HEW to set such standards within 30 days of enactment rather than the 180 days required by the administration. The requirement that each state would be made an air quality control region was also an escalation above the administrative bill. The lead time for states to adopt implementation plans was shortened, and states were allowed to set stricter standards. The feasibility clause is found on p. 921.
26. Ibid., pp. 894, 896, 901 ff. Assembly line and in-use tests were policy escalations made by the committee. Cf. also p. 916. Fuel standard discussion is found on pp. 903 f. and 932 ff. Jones, *Clean Air,* pp. 186 ff. contains a description of the pressure and discussions of the committee over fuel standards, indicating that a Congressman representing a district in which a lead company was located was responsible for the changes made.
27. *A Legislative History* (2), pp. 907 ff. Amendments were indicated that would (1) rule out the internal combustion engine by 1978, if it did not match the peak performance of alternative systems, such as steam and gas turbine; (2) require the more stringent California auto emission standards to be nationally adopted; (3) provide for voluntary federal inspection after 4,000 miles, (if an emission control device were found defective, it would be

corrected at the manufacturer's expense); and (4) restore the administration's proposals concerning fuel regulation.
28. Ibid., pp. 803, 847, 874, 878 (Mr. Staggers, committee chairman); 820 (Mr. Rogers of Florida, the leading subcommittee proponent).
29. Ibid., pp. 882 f. (Mr. Hechler); 878 (Mr. Bell of California); 885 f. (Mr. Ryan). Amendments from the deviating committee members, pp. 840 ff. (Mr. Tiernan), 849 (Mr. Van Deerlin), 875 (Mr. Farbstein), 884 (Mr. Button). See also esp. p. 885 (Mr. Ryan, amendment to strike out "feasibility" clauses).
30. Cf. Jones, *Clean Air*, pp. 191 ff. for a descriptive analysis of how action in the Senate proceeded in spring and summer 1970.
31. *A Legislative History* (1), pp. 401 ff., 410 f. The committee defined an ambient air quality standard as "the maximum permissible ambient air level of an air pollution agent . . . which will protect the health of any group of the population"; report on S. 4358.
32. Ibid., pp. 402 f., 412 f.
33. Ibid., pp. 408 ff., 414. The committee recommended special emergency powers for the responsible federal authority to provide for immediate action whenever air pollution reached such levels that it produced significant health effects, such as incapacitating or irreversible body damage in any significant portion of the population (p. 435 f).
34. Ibid., pp. 416 ff. To speed up technology development and transfer, and to prevent competitive disadvantages, the committee established a procedure of mandatory licensing that would make available, to "any person who must have access" to them, any patent, trade secret or know-how necessary to comply with the provisions of the act, p. 442 f. New source standards would be adjusted as new technology was deemed available, p. 417.
35. Ibid., pp. 424 ff., 459 f. The one-year extension could be granted only if the responsible federal authority determined after a hearing that (1) technology was not available or had not been available for a sufficient period of time, (2) the applicant had made a bona fide effort to meet the deadline, and (3) that an extension was necessary "for the general welfare of the United States," p. 427 f. Senator Dole recommended that the extension decision should be made not by the federal agency but by the Congress after receiving a recommendation from the responsible agency, pp. 447 ff. Senator Gurney suggested (1) that the Committee proposal for judicial review of the extension decision should be deleted and (2) that the federal agency should have full power to make two one-year extensions, p. 450 f.
36. Ibid., pp. 428 ff., 432 ff. The committee carefully pointed out that "control" would give more flexibility than "prohibit" to the federal agency's implementation of the provisions, p. 434. Of the $1,175,000 authorized through fiscal year 1973, 38.3 percent were for research on fuels and vehicles, pp. 472, 528.
37. Ibid., pp. 223 ff.; quot., p. 229.
38. Ibid., pp. 299 ff; quot., pp. 301, 305.
39. Ibid., pp. 392 f. The Senate vote was 73-0. The Dole and Gurney

amendments were voted down, pp. 319 and 321. One amendment provided for standing consulting committees to help the responsible federal agency get adequate information on control technology for each air pollution agent or combination of such agents, pp. 341 ff. (Senator Randolph). Another amendment provided for yearly federal agency reports on progress in auto emission control development, pp. 347 ff. (Senators Muskie and Cooper). Ibid., pp. 151 ff. (conference report); pp. 111 ff. (House debate); and 123 ff. (Senate debate). Cf. Jones, *Clean Air*, pp. 205 ff., for an account of the pressures and coverage surrounding the conference, as well as an account of its deliberations.

Chapter 4
1. SOU 1966:65 *Luftförorening, buller och andra immissioner* (Stockholm: Ministry of Justice, 1966), pp. 202 ff.; quot., p. 209.
2. Ibid., pp. 240 ff.
3. Ibid., pp. 256 ff.
4. The NEPB proposal is presented in cabinet proposal 1969:28, pp. 82 ff. See esp. p. 88. The NEPB board of directors was divided on this issue. The industrial representative, Mr. Eidem, rejected the proposal on the ground that it did not satisfy the demand for due process and the rule of law. Another member of the board, Mr. Fälldin (later to become the leader of the Center party), rejected it because of the unsatisfactory investigation preceding the proposal; cf. p. 91 f.
5. Cabinet proposal 1969:28, pp. 93 ff. Only five of the commenting agencies, organizations, and groups favored the NEPB solution; twenty-one argued for administrative licensing through an independent board while thirteen still clung to the idea of giving the Water Courts exclusive jurisdiction.
6. Ibid., pp. 195 ff. See esp. p. 200 f. According to the minister, the franchise board would deal with the most controversial and most complicated issues concerning the permissibility of polluting activities. This necessitated an air of impartiality. The environmental expert could sometimes be the NEPB director-general or someone appointed by him; this would increase the possibilities for consistency with policy developments. The minister also foresaw that as resources and experience at the regional level increased, certain exemption responsibilities could be delegated to the state regional boards. In fact, of the thirty-eight factories and other establishments and the five types of waste water covered by the franchise and exemption system in 1972, the state regional boards had the authority to decide on exemptions in twelve. It must also be mentioned that twenty-five types of factories, plants, and other establishments could not be constructed without advance notice and application to the state regional boards. Cf. Royal Ministry for Foreign Affairs et al., *Environment protection act. Marine dumping prohibition act. With commentaries. Information to the United Nations conference on the human environment* (Stockholm: ALLF 1972), pp. 69 ff.
7. Cabinet proposal 1969:28, pp. 223 ff., 236. As can be seen in chapter 5, the

real estate courts may issue a verdict containing the conditions under which a polluting activity may be continued and even prohibit such activity, provided the plaintiff's case concerns an activity for which there exists only an exemption decision. As soon as there exists a permit from the FBEP, the courts only have jurisdiction to try cases of compensation.
8. Ibid., pp. 366 ff. (Cabinet's Legal Advisory Council); Motions 1969 I:947 and II:1083 (Center party), I:946 and II:1086 (Liberal party), and I:945 and II:1085 (Conservatives); Committee Report, LU³ 1969:37, pp. 163 ff. (minority view of the three bourgeois parties); *Riksdag Record* AK 1969 28:64 (Mr. Grebäck, Center); 28:67 f. (Mr. Tobé, Liberal).
9. Cabinet proposal 1969:28, p. 394 (the minister); Committee Report LU³ 1969:37, p. 147 f. (Social-Democratic majority view); *Riksdag Record* AK 1969 28:61 (Mr. Kling, minister of justice); 28:73 (Mr. Svenning, Social Democrat). The votes were 72–60 in the First Chamber, and 115–97 in the Second Chamber for the cabinet proposal.
10. Motions 1969 I:868 and II:482, I:946 and II:1086 (Liberal).
11. Committee Report LU³ 1969:37, p. 142 (NEPB comment); p. 157 f. (Social-Democratic majority view); p. 171 f. (Liberal-Center minority view). *Riksdag Record* AK 1969 28:71 (Mrs. Anér, Liberal). The margin was larger than 2 to 1 in each chamber.
12. *A Legislative History* (2), pp. 1550 ff. The enforcement procedure involved a *conference,* followed by abatement recommendations by the Secretary of HEW, and, if no adequate abatement was forthcoming, a public *hearing* before a special panel. If abatement was not forthcoming following the recommendations of the panel, the federal government could initiate *court action.* This, however, concerned interstate pollution. Federal court action against intrastate pollution must await a request by the state governor. With a specified time period of six months between the issuing of recommendations and the initiation of the next step in the procedure, this process was indeed "time-consuming and cumbersome."
13. Ibid., pp. 1503 ff. (President's message); pp. 1476, 1479, 1483 ff. (S. 3466, the administration bill). See esp. the summary, pp. 1495 ff.
14. Ibid., pp. 891 ff. (House Committee report). For an analysis of the organizational problems at the federal level, see Charles O. Jones, "The Limits to Public Support: Air Pollution Agency Development," *Public Administration Review* 33 (1972): 502 ff.
15. *A Legislative History* (2), pp. 891 ff.; 911 ff.
16. Ibid., pp. 799 (Mr. Ryan); 814 (Mr. Skubitz, quote); 815 (Mr. Jarman); 837 (Mr. Vanik, quote); 844 (Mr. Monogan, quote); 882 (Mr. Hechler of W. Va., quote); 803 ff. (Mr. Staggers, quote).
17. Ibid., pp. 857 ff. (Mr. Saylor); cf. p. 861 f. (Mr. Mikva).
18. Ibid., pp. 859 f. (Mr. Springer); 860 (Mr. Rogers); 869 f. (Mr. Dingell); 871 (Mr. Staggers).
19. *A Legislative History* (1), pp. 409 ff., 454 ff., 486 (ambient air quality standards and goals); 415 ff., 457 f., 491 ff. (new source performance standards, emission standards for selected agents and hazardous pollutants).

208 *Notes*

20. Ibid., pp. 411 ff., 455 f., 486 ff. (state implementation plans); 421 ff., 458 f., 497 f. (federal enforcement); 405 f., 452, 472 ff. (federal grants to pollution control programs).
21. Ibid., pp. 415, 432 (state standards); pp. 411 ff., 455 f., 486 ff. (implementation plans); 417 (delegation of power to states to issue certificates for new sources).
22. Ibid., pp. 440 ff., 465 f., 525 (judicial review of administrative standards and guidelines); 428, 460 f., 503 f. (judicial review of suspension decisions); 447 ff. (Dole proposal for Congressional power to make suspension decisions).
23. Ibid., p. 260 (Senator Cooper); pp. 410 ff., 421 ff. (committee report). Cooper said the earlier concept of national emission standards "raised great problems of fairness and federal determination of local consequences," and did not assure achievement of clean air.
24. Ibid., pp. 261, 309 f., 386 (Cooper); 265 f., 311 f. (Baker); 271 f., 295 ff., 309, 314 (Dole); 292 f., 315 (Gurney); 228, 299, 320 (Muskie); 320 (Randolph). For the votes, see p. 319 (Gurney amendment), and 321 (Dole amendment).
25. "Message of the President Relative to Reorganization Plans Nos. 3 and 4 of 1970, July 9, 1970," *Environmental Quality. The First Annual Report of the Council on Environmental Quality* (Washington, D.C.: Govt. Printing Office, 1970), pp. 294 ff.
26. Jones, *Clean Air,* p. 206.
27. *A Legislative History* (1), pp. 152 f., 156 ff., 162 ff. (Conference bill); pp. 193 ff. (statement on the part of the managers of the House).
28. Ibid., p. 116 (Springer); 137 (Muskie); 260 (Cooper).

Chapter 5

1. The literature on techniques for public participation is large. For some applications to environmental and natural resource policies, see Albert E. Utton et al., *Natural Resources for a Democratic Society. Public Participation in Decision-Making* (Boulder, Colo.: Westview Press, 1976).
2. The classic arguments for public participation are listed in Herbert McClosky, "Political Participation," *International Encyclopedia of the Social Sciences* (1968), vol. 12, pp. 252 ff.
3. Cabinet proposal 1967:59, pp. 53 ff.; motions 1967 I:770 and II:958 (Center Party); Committee Report JoU 1967:17, pp. 8f., 22 f. The Air Quality Council in 1974 consisted of ten members, of which three represented science, two business, two local communities, and three professional interests (meteorology, architecture, law); cf. *Naturvårdsverkets årsbok 1974* (Stockholm: ALLF, 1975), pp. 116f. Between 1969 and 1975, the board of directors of the NEBP consisted of the director-general, one parliamentarian (Center Party), one representative of local community interests, one for business, two for consumer groups, and one legal expert.
4. SOU 1966:65, *Luftförorening, buller och andra imissioner* (Stockholm: Ministry of Justice, 1966), pp. 225 ff.
5. Lundqvist, *Miljövårdsförvaltning,* pp. 192 ff.

6. SOU 1966:65, *op. cit.*, pp. 299, 307 f.; cabinet proposal 1969:28, pp. 189 ff. (minister), 375 ff. (Legal Advisory Council); 205 f., 219, 395 f. (minister).
7. Cabinet proposal 1969:28, pp. 203 ff., 285 f.
8. Ibid., pp. 370 f. (Legal Advisory Council); 394 ff. (minister).
9. Motions I:947 and II:1083 (Center Party); I:945 and II:1085 (Conservatives).
10. Motions I:946 and II:1086 (Liberal Party).
11. The three opposition parties supported changes that would (1) guarantee compensation to affected parties for the costs of partaking in FBEP hearings, (2) guarantee compensation to affected parties for their costs of partaking in real estate court proceedings; cf. committee report LU^3 1969:37, pp. 166 ff.; the Liberals and the Center Party supported an amendment concerning the right for local communities to appeal FBEP decisions; cf. ibid., p. 169. For the debate, see the *Riksdag Record*, AK 1969 28:89 f. (Mr. Hugosson, Social Democrat), and 28:83 (Mrs. Anér, Liberal) (italics added).
12. *A Legislative History* (2), pp. 1474 ff. (administration bill, S. 3466); pp. 910 ff. (House bill, H.R. 17255); pp. 891 ff. (House report).
13. Ibid., vol. 1, pp. 412, 419, 431 f. (Senate report).
14. Ibid., pp. 441 f., 465 f., 525 (Senate report).
15. Ibid., pp. 704 ff. (Senate Committee on Public Works, Print No. 1, August 25, 1970); pp. 436 ff., 464 f., 522 f. (Senate report, September 17, 1970).
16. Ibid., pp. 717 (American Mining Congress); 724 f. (Automobile Manufacturers Association); 748 (Ford Motor Company); 754 (Manufacturing Chemists Association); 782 (Standard Oil of Indiana); 788 (Union Carbide Corporation). One change in the legislation should be noted. While the first print of August 25 stated only that suits may be brought by "one or more persons," the final version stated that suits may be brought by "one or more persons on their own behalf." This qualification was present also in S. 4358. Cf. ibid., pp. 704 (Committee Print No. 1); 613 (S. 4358); 522 (Senate report).
17. Ibid., pp. 230 (Senator Muskie); 263 (Senator Spong); 273 ff. (Senator Hruska's criticism).
18. Ibid., pp. 280, 351 ff. (Muskie).
19. A "class action" involves suits for recovery of damage to an identifiable class of citizens, and is not primarily concerned with seeking enforcement of a certain provision in a particular law. In the words of the Senate committee report: "Questions with respect to traditional 'class' actions often involve: (1) identifying a group of people whose interests have been damaged; (2) identifying the amount of total damage to determine jurisdiction qualification; and (3) allocating any damage recovered. None of these points is appropriate in citizen suits seeking abatement of violations of air quality standards"; ibid., p. 438. However, cf. p. 464. For a thorough discussion of the citizen suits provision, see Richard E. Ayres and James F. Miller, *Citizen Suits Under the Clean Air Act* (Washington, D.C.: U.S. Environmental Protection Agency, n.d.).
20. *A Legislative History* (1), pp. 349 (Senator Griffin); 354 (Senator Cook); 351, 353 (Muskie; italics added).

21. Ibid., pp. 211 ff. (administration's letter to conference committee recommending certain provisions). See esp. p. 214 f.
22. Ibid., pp. 182 ff., 205 f. (conference report); 112, 117 (House debate, Representatives Staggers and Springer); 138 (Senate debate, Senators Eagleton and Muskie). According to Ayres and Miller, *Citizen Suits*, another limitation is present in the final legislation: a citizen may not sue in a state in which he does not reside.

Chapter 6

1. *A Legislative History* (1), pp. 401, 411 (committee report); 224 f. (Senator Muskie).
2. Swedish cabinet proposal 1969:28, p. 179 f. Cf. cabinet proposal 1968:160, pp. 32 ff.
3. Royal Ministry for Foreign Affairs et al., *Air pollution across national boundaries. The impact on the environment of sulfur in air and precipitation. Sweden's case study for the United Nations conference on the human environment* (Stockholm: Norstedts, 1971), passim.
4. Estimates of lead pollution are based on figures in Royal Ministry for Foreign Affairs et al., *Sweden's national report to the United Nations on the human environment* (Stockholm: Norstedts, 1971), p. 35; and John Quarles, *Cleaning Up America. An Insider's View of the Environmental Protection Agency* (Boston: Houghton Mifflin Company, 1976), p. 119 f.
5. SOU 1967:43, *Miljövårdsforskning. Del I:Forskningsområdet* (Stockholm: Ministry of Agriculture, 1967), pp. 28 ff.; SOU 1970:13, *Sveriges energiförsörjning. Energipolitik och organisation* (Stockholm: Ministry of Industry, 1970), pp. 20 ff., 66; *Statistisk Årsbok för Sverige 1973* (Stockholm: SCB, 1973), p. 177.
6. Energy projections are from James Rathlesberger (ed.), *Nixon and the Environment. The Politics of Devastation* (New York: Village Voice/Taurus, 1972), pp. 81 ff., 90 ff., 96. See also *Statistical Abstracts of the United States 1970* (Washington, D.C.: U.S. Department of Commerce, 1970), pp. 505 ff., 545, and 821 f.
7. *A Legislative History* (2), pp. 803 (Mr. Staggers), 896 (Committee Report); ibid., (1), pp. 239 f. (Senator Muskie).
8. Cf. Jones, *Clean Air*, p. 175 f.
9. Swedish cabinet proposal 1968:122, pp. 12 ff.; cabinet proposal 1968:160, pp. 32 ff. (Minister of Transport, Olof Palme); cf. Chapter 3, note 9.
10. *A Legislative History* (1), p. 403 (Senate committee report).
11. Ibid., pp. 224, 231 (Senator Muskie).
12. Swedish cabinet proposal 1969:28, p. 216 (Minister of Justice Herman Kling).
13. SOU 1970:13, op. cit., p. 67. These figures had already been communicated to the public and to policymakers in 1968.
14. Paul H. Gerhardt, *An Approach to the Estimation of Economic Losses Due to Air Pollution* (Washington, D.C.: National Air Pollution Control Administration, 1968).
15. *Statistisk Årsbok för Sverige 1968* (Stockholm: SCB, 1968), p. 29; SOU

1966:69, *Trafikutveckling och trafikinvesteringar* (Stockholm: Ministry of Finance, 1966), pp. 16, 48. There were 231 cars per 1,000 population in 1965.
16. *Statistical Abstracts of the United States 1970*, pp. xiii, 506 f., 535.
17. Cabinet proposal 1969:1, Section 11, p.169 f.
18. *A Legislative History* (1), pp. 227, 232 (Senator Muskie).
19. Ibid., pp. 300 f., 308 (Senator Griffin).
20. In a sense, this view is consistent with the one taken by David R. Mayhew, *Congress: The Electoral Connection* (New Haven, Conn.: Yale University Press, 1974), which conjures up a vision of United States congressmen as "single-minded seekers of reelection"; p. 5.
21. Charles O. Jones, "Why Congress Can't Do Policy Analysis (or words to that effect)," *Policy Analysis* 2 (1976): 255.
22. Heidenheimer et al., *Comparative Public Policy*, p. 262.
23. The results of the U.S. polls between 1965 and 1970 have been summed up in two articles by Hazel G. Erskine, "The Polls: Pollution and Its Costs," *Public Opinion Quarterly* 36 (1972): 120–35, and "The Polls: Pollution and Industry," ibid. (Summer 1972): 263–280.
24. Jamie Heard, "Washington Pressures/Friends of the Earth Give Environment Interests an Activist Voice," *National Journal 2* (1970): 1711 ff. For an account of the Earth Day activities in April 1970, see ibid., pp. 408 ff.
25. The Nader report was John C. Esposito's *Vanishing Air* (New York: Grossman Publishers, 1970). The criticism of Muskie's policy record is found on pp. 288 ff.
26. Cf. Jones, *Clean Air*, pp. 191, for a description of how Senate action proceeded in spring and summer 1970.
27. *A Legislative History* (1), p. 231 (Senator Muskie); p. 308 (Senator Griffin).
28. Ibid., p. 225 f. (Senator Muskie); ibid., (2), p. 895 (House Committee Report).
29. Anthony King, "Ideas, Institutions, and the Policies of Governments: Part III," *British Journal of Political Science* 3 (October 1973): 419.
30. *A Legislative History* (1), p. 309 (Senator Cooper).
31. Jones, *Clean Air*, p. 176.
32. Lundqvist, *Miljövårdsförvaltning*, pp. 105 ff.
33. For a description of the Swedish structure of government, see Donald M. Hancock, *Sweden. The Politics of Postindustrial Change* (Hinsdale, Ill.: Holt, Rinehart & Winston, Dryden Press, 1972), chap. 7–8.
34. Cf. ibid., chap. 8. See also Hans Meijer, "Bureaucracy and Policy Formulation in Sweden," *Scandinavian Political Studies* 4 (1969). 103–16 and Thomas J. Anton, "Policy-making and Political Culture in Sweden," ibid., pp. 82–102.
35. Cf. Anton, "Policy-making," pp. 95 ff.
36. *Riksdag Record* FK 1969 28:138 f.
37. Committee Report LU^3 1969:37, p. 157.
38. Cabinet proposal 1969:28, p. 395 (Minister of Justice Herman Kling).
39. This discussion combines the perspectives of Jones, *Clean Air*, p. 176;

212   Notes

Heidenheimer et al., *Comparative Public Policy*, p. 262; and Aaron Wildavsky, "The Strategic Retreat on Objectives," *Policy Analysis* 2 (1976): 520 ff.

Chapter 7

1. The legally prescribed context of policy change is found in PL 91-604, the *Clean Air Act Amendments of 1970*, Sec. 202. For the complete text, cf. *A Legislative History* (1), pp. 81 ff.
2. National Academy of Sciences, *Report by the Committee on Motor Vehicle Emissions* (Washington, D.C.: Govt. Printing Office, 1973), p. 126. It seems clear that the NAS committee was uncertain about the virtues of the catalytic converter technology opted for by major American car manufacturers. The committee seemed more inclined to favor the Japanese dual-carbureted, stratified-charge engine (Honda), which could meet the 1975-76 standards for 50,000 miles. Furthermore, the catalyst technology might impose three to four times greater costs for annual operation and maintenance than the Japanese technology, due to the serious fuel penalties associated with the catalyst technology; ibid, pp. 87 ff., 100 ff.
3. U.S., Congress, Senate, Subcommittee on Air and Water Pollution, *Hearings on the Decision of the Administrator of the Environmental Protection Agency Regarding Suspension of the 1975 Auto Emission Standards*, 93d Cong., 1st Sess. (Washington, D.C.: Govt. Printing Office, 1973; hereinafter cited as *EPA Decision Hearings*), Part 1, pp. 1-52 (quote, p. 8). "The most compelling factor" in Ruckelshaus's decision was the possibility that installing catalysts on the "entire product line . . . may well . . . cause significant economic disruption"; p. 9.
4. Cf. "Air Pollution Control Deadline Postponed," *Congressional Quarterly Almanac 1973*, pp. 653 f.
5. *EPA Decision Hearings Part 1*, p. 54 (Senator Randolph, chairman of the full Public Works Committee); p. 57 (Senator Buckley); pp. 185 ff. (Senator Muskie, quote p. 188). Ibid., *Part 2*, p. 464 (Muskie). Ibid., *Part 3*, p. 1022 (Muskie). Ford Motor Co. wanted another one-year delay; ibid., p. 1028 (Mr. Lee Iaccocca). GM wanted the present interim standards for California frozen into law for "several years"; ibid., p. 1322 (Mr. Edward N. Cole). Also, Chrysler wanted further postponement; ibid., p. 1569 (Mr. John J. Riccardo).
6. Ibid., *Part 1*, p. 57. In the hearings on June 26, 1973, Senator Randolph pointed out that the "strong steps" taken in 1970 were taken because of the "overriding concern" for the "protection of public health," but also with "the full knowledge that industry and citizens . . . would be affected in ways that were at variance with the traditional habits which we had in our economy over a long period of time." The responsibility of those who passed that legislation to see if there was a need "to pull back" had now taken on an "additional significance" because of the "gasoline shortage"; ibid., *Part 4*, p. 1689 f.
7. *Congressional Record* (daily edition), Dec. 17, 1973, p. S23084 (Senator Randolph); p. S23081 (Senator Baker); p. S23081 (Senator Muskie on possible fuel savings).

Notes 213

8. The president's message to Congress on "The Energy Crisis" is published in Council on Environmental Quality, *Environmental Quality. The Fifth Annual Report of the Council on Environmental Quality* (Washington, D.C.: Govt. Printing Office, 1974), pp. 551 ff. For the proposed changes in the Clean Air Act, see p. 560 (quote).
9. U.S., Congress, Senate, *Energy Emergency Act, Conf. Rept. S 681 to accompany S. 2589*, 93d Cong., 2d Sess., 1974, pp. 97 f. *Congressional Record* (daily edition), Feb. 19, 1974, p. S3467; ibid., Feb. 27, 1974, p. H4439. Muskie and the other sponsors of the original Clean Air Act had to fight on two fronts. Oil state Senators and Representatives from rural districts wanted further delays and rollbacks in clean air laws; the Wyman amendment in the House, calling for a complete suspension of *all* emission control requirements for 90 percent of the nation's area until 1977, was narrowly defeated in the House in December, 1973; *Congressional Quarterly Almanac 1973*, p. 682. More environmentalist-oriented policymakers accused Muskie and others of "using the energy crisis as an excuse [for] sweeping aside" environmental safeguards; *Congressional Record* (daily edition), February 19, 1974, pp. S3431 ff. and S3442 (Senators Percy and Weicker). Muskie and Baker defended the clean air provisions as a reasonable compromise between health, fuel economy, and auto industry viewpoints, which reflected "a pragmatic understanding of political realities" (Baker); ibid., January 29, 1974, pp. S1128, S1139, S1151. Cf. ibid., February 19, 1974, pp. S3431 ff. (Muskie).
10. U.S., Congress, Senate, *Veto Message from the President of the United States* returning (S. 2589) The Energy Emergency Act, S Doc. 61, 93d Cong., 2d Sess., March 6, 1974.
11. U.S., Congress, House, *Energy Supply and Environmental Coordination Act of 1974, H. Rept. 1013 to accompany H.R. 14368*, 93d Cong., 2d Sess., 1974, pp. 7, 53. *Congressional Record* (daily edition), May 1, 1974, p. H12539.
12. "Congress Votes to Delay Clean Air Standards," *Congressional Quarterly Almanac 1974*, p. 743. *Congressional Record* (daily edition), May 14, 1974, p. S14546.
13. U.S., Congress, House, *Energy Supply and Environmental Coordination Act, Conf. Rept. H 1085 to accompany H.R. 14368*, 93d Cong., 2d Sess., 1974, pp. 14, 40 f. *Congressional Quarterly Almanac 1974*, p. 738.
14. U.S., Congress, House, *Rept. H 1085*, pp. 13 ff.
15. U.S., Congress, House, *Congressional Record* (daily edition), May 1, 1974, pp. H12511 (Mr. Wyman); H12529 (Mr. Jarman); H14533 (Mr. McEwen).
16. Ibid., pp. H12514, H12526 f., H12532 (Mr. Rogers); H12517 (Mr. Broyhill); H12529 (Mr. Nelsen); H12534 (Mr. Staggers).
17. U.S., Congress, Senate, *Congressional Record* (daily edition) May 14, 1974, pp. S14527 f. (Muskie); S14532 (Randolph).
18. Cf. "Congress Faces Hard Choices on Clean Air Act," *Congressional Quarterly Almanac 1975*, p. 249 f. President Ford, however, did ask the auto industry to improve fuel economy by 40 percent from 1974 to 1980 models.

19. Cf. Arthur J. Magida, "EPA Study May Bring Reprieve for the Catalytic Converter," *National Journal Reports* 7 (1975), pp. 552 ff.
20. U.S., Congress, Senate, *Clean Air Act Amendments of 1976, S. Rept. 717 to accompany S. 3219*, together with minority and individual views, 94th Cong., 2d Sess., 1976, pp. 5, 56 ff. The committee emphasized EPA findings that it was technically feasible "to achieve any of the currently legislated emission standards," p. 59. For Senator Muskie's individual views, see p. 99.
21. U.S., Congress, House, *Clean Air Act Amendments of 1976, H. Rept. 1175 to accompany H.R. 10498*, together with additional, separate, opposing, and minority views, 94th Cong., 2d Sess., 1976, pp. 195 ff. In an additional view, Paul Rogers said the legislation would achieve the goal of balancing "our health requirements with other national needs—economic, energy and growth"; ibid., p. 413. The antienvironmentalist arguments are found in the dissenting views of Dingell and others, pp. 415 ff. Cf. "Committee Markup: Auto Emission Standards," *Congressional Quarterly Weekly Report* 34 (1976): 574, for the Brodhead and Rogers comments.
22. U.S., Congress, Senate, *Congressional Record* (daily edition), July 26, 1976, pp. S12478 f. (Muskie); S12461 (Buckley); S12465 f. (Baker); S12487 f. (Muskie). Here, Muskie stated that it was *not* technical feasibility but, rather, economic recovery and fuel economy that dictated the committee's standpoint on the relaxation of standards.
23. U.S., Congress, Senate, *Congressional Record* (daily edition), August 5, 1976, pp. S13487 ff. (Gary Hart). For a supporting view, see p. S13499 (Cranston). For a dissenting view on the Hart amendment, see p. S13497 (Griffin).
24. Ibid., p. S13499 (Muskie). For the roll-calls on the Hart amendments, see pp. S13498 and S13500.
25. U.S., Congress, House, *Congressional Record* (daily edition), August 4, 1976, pp. H8294 ff. (Rogers); H8321 ff. (Waxman); H8330 f. (Maguire); H8332 (Drinan: the energy crisis is a "godsend to those who want to delay compliance to clean air standards").
26. Ibid., pp. H8307 ff. (Dingell). It should be noted that Rep. Dingell represents Dearborn, Mich., in the heart of the auto-manufacturing region. Cf. p. H8299 (Broyhill).
27. Ibid., September 15, 1976, p. H10089 f.
28. Cf. "Clean Air Agreement," *Congressional Quarterly Weekly Report* 34 (1976): 2725; Prudence Crewdson, "Clean Air Amendments Die at Session's Close," ibid. (1976): 2927 ff.
29. For a summary of the committee action on HR 6161, see James R. Wagner, "House Weakens Clean Air Amendments," *Congressional Quarterly Weekly Report* 35 (1977): 1028 ff. The Dingell-Broyhill substitute lost in committee on a tie vote April 27, after a bitter debate. Nine Democrats and nine Republicans entered a forty-page "separate view" in the committee report, promising to offer the Dingell-Broyhill bill as an amendment on the House floor.
30. Cf. ibid., pp. 1023 ff. for a summary of the House floor action on the 1977

Clean Air Act Amendments. The strong industry-union support for the Dingell-Broyhill amendment is indicated on pp. 1024 f.
31. For a summary of the committee action on S 252, see James R. Wagner, "Controversial Clean Air Report Filed,"*Congressional Quarterly Weekly Report* 35 (1977): 960 ff. The Senate floor action is summed up by Wagner in "Senate Adopts Clean Air Compromise," ibid., pp. 1135 ff.
32. The premises for the conference work, as well as the outcome of the conference deliberations, are found in James R. Wagner, "Senate-House Differences Promise Tough Conference on Clean Air Amendments," *Congressional Quarterly Weekly Report* 35 (1977): 1223; "Conference Outlook: Clean Air Act Amendments," ibid., p. 1379; and James R. Wagner, "Clean Air Bill Extends Deadlines," ibid., pp. 1629 ff. See also J. Dicken Kirschten, "The Clean Air Conference: Something for Everybody," *National Journal* 9 (1977): 1263 ff.
33. The legally prescribed context of policy implementation and change is found in Sec. 110 of the *1970 Clean Air Act Amendments*. For the complete text, see *A Legislative History (1)*, pp. 14 ff.
34. The implications for land use and growth, and for energy, are discussed in Council for Environmental Quality, *Environmental Quality. The Fourth Annual Report of the Council on Environmental Quality* (Washington, D.C.: Govt. Printing Office, 1973), pp. 159 ff. In 1973, EPA was forced by court order to issue regulations requiring states to approve in advance the siting and constructions of both new polluting facilities and such complex facilities as shopping centers, which could attract concentrations of vehicles and thereby lead to violation of ambient standards. The report also stated that domestic low-sulfur fuel supplies "are inadequate to meet the increased demand resulting from the SO_2 control approach of many state implementation plans." In its *Fifth Annual Report* of 1974 (cf. note 8 for this chapter), the CEQ refers to the controversy between the utility companies and the EPA concerning the availability of stack gas scrubbers. In March 1974, the EPA tried unsuccessfully to have Congress pass an amendment permitting the indefinite use of intermittent control systems; cf. *Fifth Annual Report*, pp. 121 ff.
35. Council on Environmental Quality, *Environmental Quality 1973*, p. 162. Cf. *Environmental Quality 1974*, pp. 118 f. New York and Philadelphia metropolitan areas required the use of 0.3 percent sulfur oil, Connecticut and Boston area 0.5, and Rhode Island, Maryland, and the remainder of New Jersey 1.0 percent.
36. U.S., Congress, House, *Energy Supply and Environmental Coordination Act of 1974, H. Rept. to accompany H.R. 14368*, pp. 14, 17 ff.
37. U.S., Congress, Senate, *Congressional Record* (daily edition), May 14, 1974, pp. S14523 ff.; the Senate committee reported by way of inserting Muskie's substitute directly in the *Record* at the beginning of the debate.
38. Ibid., pp. S14526 ff. (Muskie); see esp. p. S14528.
39. Ibid., pp. S14531 ff. (Randolph); see esp. pp. S14534 and S14536.
40. U.S., Congress, House, ibid., May 1, 1974, pp. H12513 ff. (Abzug and others); U.S. Congress, House, *Energy Supply and Environmental*

Coordination Act of 1974, Conf. Rept. H 1085 to accompany H.R. 14368, pp. 31 ff. Cf. *Congressional Quarterly Almanac 1974*, p. 744.

41. U.S., Congress, Senate, *Congressional Record* (daily edition), May 14, 1974, p. S14527 (Muskie) and S14532 (Randolph).
42. As a prelude to the major revision of the 1970 act, due in 1975, the relevant subcommittees of both houses held series of hearings in 1974 and 1975. The 1970 act was extended until mid-1976, since no reports came out of committee until spring of 1976. For an overview of the "great scrubber debate," see the following: "Problems Cited in Meeting Clean Air Deadlines," *Congressional Quarterly Weekly Report* 32 (1974): 1385 f.; *Congressional Quarterly Almanac 1974*, p. 739 f.; and *Congressional Quarterly Almanac 1975*, p. 248 f. See further John F. Burby, "EPA Alive and Well, but Meeting Stiffer Resistance," *National Journal Reports* 6 (1974): 431 ff.; James A. Noone, "Great Scrubber Debate Pits EPA Against Electric Utilities," ibid., (1974), pp. 1103 ff.; and Council on Environmental Quality, *Environmental Quality. The Sixth Annual Report of the Council on Environmental Quality* (Washington, D.C.: Govt. Printing Office, 1975), pp. 46 ff.
43. John F. Burby, "New Committee Lineups Will Shape Clean Air Act Revisions," *National Journal Reports* 7 (1975): 13.
44. *Congressional Quarterly Almanac 1975*, p. 247 f.
45. U.S., Congress, House, *Clean Air Act Amendments of 1976, H. Rept. 1175 to accompany H.R. 10498*, p. 409 (additional views of Paul Rogers, Chairman, Subcommittee on Health and Environment).
46. U.S., Congress, Senate, *Clean Air Act Amendments of 1976, S. Rept. 717 to accompany S. 3219*, pp. 34, 42 f., 99 (Muskie's individual view).
47. U.S., Congress, Senate, *S. Rept. 717*, pp. 7 ff., 34 ff., 41 ff.; U.S., Congress, House, *H. Rept. 1175*, pp. 40 ff., 168 ff.
48. U.S., Congress, Senate, *S. Rept. 717*, p. 42; U.S., Congress, House, *H. Rept. 1175*, pp. 179 ff.
49. U.S., Congress, Senate, *Congressional Record* (daily edition), July 26, 1976, pp. S12462 (Buckley); S12483 (Muskie); ibid., August 5, 1976, pp. S13527 f. (Morgan); S13528 f. (Allen).
50. Royal Ministry for Foreign Affairs et al., *Environment protection act. Marine dumping protection act. With commentaries.* Information to the United Nations conference on the human environment (Stockholm: ALLF, 1972), p. 54 f.
51. U.S., Congress, Senate, *Congressional Record* (daily edition), August 5, 1976, pp. S13528 (Muskie); S13529 and S13536 ff. (Randolph).
52. U.S., Congress, House, ibid., September 15, 1976, pp. H10121 ff. The motion to recommit was put forward by Mr. Broyhill.
53. U.S., Congress, House, *Clean Air Act Amendments of 1976, Conf. Rept. H 1742 to accompany S. 3219*, 94th Cong., 2d Sess., 1976, pp. 88, 90, 93 ff. Cf. *Congressional Quarterly Weekly Report* 34 (1976): 2927 ff.
54. *Congressional Quarterly Weekly Report* 35 (1977): 960 ff.
55. Ibid., pp. 1023, 1027 f.
56. Ibid., pp. 1629 ff. See also Kirschten, "The Clean Air Conference—Something for Everybody," *National Journal* 9 (1977): 1261 f.

57. James R. Wagner, "President Signs Revisions in 1970 Clean Air Law," *Congressional Quarterly Weekly Report* 35 (1977): 1713 ff.
58. U.S., Congress, Senate, *Congressional Record* (daily edition), July 26, 1976, p. S12478. Cf. James A. Noone, "Scientists' Study Upholds Clean Air Standards," *National Journal* 6 (1974): 1389 f.
59. U.S., Congress, House, *Clean Air Act Amendments of 1976, H. Rept. 1175 to accompany H.R. 10498*, p. 32 f.
60. U.S., Congress, Senate, *Clean Air Amendments of 1976, S. Rept. 717 to accompany S. 3219*, p. 42.
61. Quarles, *Cleaning Up America*, p. 205. He continues: "This was to remain one of the most widespread irritants eating away on the basis of public support for environmental programs. . . . Complaints were especially strong in many rural areas not plagued by air pollution, since the need for the controls in those areas was difficult to defend."
62. U.S., Congress, Senate, *Congressional Record* (daily edition), July 26, 1976, p. S12478 f.
63. Quarles, *Cleaning Up America*, p. 213 f.
64. Ibid., p. 214 (contains the quotation from Muskie).
65. Riley E. Dunlap and Kent D. Van Liere, "Further Evidence of Declining Public Concern with Environmental Problems: A Research Note," *Western Sociological Quarterly* 8 (1977): 108 ff., contains a discussion of their own as well as others' research on environmental opinion.
66. Neeltje Wiedemeyer, "The Polls: Do People Worry About the Future?" *Public Opinion Quarterly* 40 (1976): 382.
67. Connie de Boer, "The Polls: Nuclear Energy," *Public Opinion Quarterly* 41 (1977): 402, 405. In a contracted EPA study of June 1974, *Impact of the Fuel Shortage on Public Attitudes toward Environmental Protection* (Washington, D.C.: EPA, June 1974), p. 26, the following figures appear concerning the perceived usefulness and favorability ratings of items suggested to cope with the fuel shortage:

	Usefulness "A Lot/Some"	Favorability "For"
Relax auto emission schedules	41	46
Allow the burning of coal and other less clean fuels	46	42
Take antipollution devices out of cars	49	41
Relax pollution controls for factories	38	31

68. Quarles, *Cleaning Up America*, p. 208 f.
69. U.S., Congress, Senate, *Congressional Record* (daily edition), July 27, 1976, p. S12561.
70. Cf. James A. Noone, "Great Scrubber Debate Pits EPA Against Electric Utilities," *National Journal Reports* 6 (1974): 1103 ff.; and by the same author, "Energy Issues Threaten Recent Environmental Gains," ibid., pp. 305 ff.; see also John F. Burby, "EPA Alive and Well, But Meeting Stiffer Resistance," ibid., pp. 431 ff.

218 Notes

71. Prudence Crewdson, "Clean Air Lobbying: Non-Deterioration," *Congressional Quarterly Weekly Report* 34 (1976): 1035 ff.; James R. Wagner, "House Weakens Clean Air Amendments," ibid., 35 (1977): 1023 ff.
72. James R. Wagner, "Senate Adopts Clean Air Compromise." *Congressional Quarterly Weekly Report* 35 (1977): 1135. The lobbying continued throughout the conference deliberations on the bill, see Wagner, "Clean Air Bill Extends Deadlines," ibid., p. 1630.
73. Quarles, *Cleaning Up America*, pp. 201 ff. For an account of the relations between EPA and the White House, see his chap. 7.
74. Cf. Burby, "New Committee Lineups Will Shape Clean Air Act Revisions," *National Journal Reports* 7 (1975): 13. See also Arthur J. Magida, "Clean Air Act Deliberations—The Changing of the Guard," ibid., 8 (1976): 340 f.
75. For examples of an overview of this protracted process, see Prudence Crewdson, "Congress Faces Hard Choices on Clean Air Act," *Congressional Quarterly Weekly Report* 33 (1975): 1169 ff.; "Committee Consideration: Clean Air Act Amendments," ibid., p. 1511; "Committee Action: Auto Emissions," ibid., p. 1638; and "Energy and Environment Notes: Clean Air," ibid., p. 2238. See further Richard Corrigan, "Congress Is Gearing Up for Major Fight on the Clean Air Act," *National Journal* 8 (1976): 901 f.; Arthur F. Magida, "Clean Air Act Deliberations—The Changing of the Guard," ibid., p. 340 f.; Prudence Crewdson, "Senate Committee Action: Clean Air Amendments," *Congressional Quarterly Weekly Report* 34 (1976): 311; "Committee Markup: Auto Emission Standards," ibid., p. 574. See also U.S., Congress, Senate, *Congressional Record* (daily edition), July 26, 1976, p. S12456 (Senator Randolph, comments on the Committee's work); U.S., Congress, House, ibid., August 4, 1976, p. H8294 (Paul Rogers, similar comments); and finally J. Dicken Kirschten, "It's Washington Taking on Detroit in the Auto Pollution Game," *National Journal* 9 (1977): 9ff., esp. pp. 12 f.
76. On the notion of "subgovernments," with a special bearing on environmental policymaking, see Cynthia H. Enloe, *The Politics of Pollution in a Comparative Perspective* (New York: David McKay Co., 1975), pp. 168 ff.
77. *Congressional Quarterly Weekly Report* 34 (1976): 2928.
78. Kirschten, "It's Washington Taking on Detroit," p. 15.

Chapter 8

1. According to statistics in the NEPB Yearbooks, the NEPB decided more than 1,200 exemption cases between 1969 and 1973. During the same time, the Franchise Board issued more than 350 permit decisions, and the State Regional Boards almost 850 exemptions. After 1975, between 30 and 40 percent of the NEPB decisions concerned changes in earlier exemption decisions. Cf. NEPB, *Naturvårdsverkets årsbok 1972* (Stockholm: ALLF, 1972), pp. 164f.; ibid., *1973*, pp. 141 ff.; ibid., *1976–1977*, pp. 156 ff. No wonder then that the NEPB was beginning to rethink the individual source control approach by the mid-1970s; cf. below, note 3.

2. See SOU 1966:65, *Luftföroreningar, buller och andra immissioner* (Stockholm: Ministry of Justice, 1966), p. 229, and cabinet proposal 1969:28, p. 219.
3. NEPB, *Riktlinjer för luftvård* (Stockholm: NEPB, 1973:8).
4. NEPB, *Naturvårdsverkets årsbok 1973* (Stockholm: ALLF, 1973), p. 77 f.
5. Ulf Högström, "Air Pollution in Swedish Communities," Ambio 4 (1975): 120 ff. Högström's article is based on a 1974 report prepared "at the request" of the NEPB "to supply data for discussion of Air Quality Criteria," p. 124.
6. P. Camner et al., "Air Quality Criteria and Guides for Sweden in Regard to Sulfur Dioxide and Suspended Particulates," *Nordisk Hygienisk Tidskrift* (1973), Suppl. 5. The relevant WHO document is *Air Quality Criteria and Guides for Urban Air Pollutants* (Geneva: WHO Techn. Rep. Ser. No. 506, 1972).
7. NEPB, *Riktvärden för luftkvalitet–svaveldioxid och stoft* (Stockholm: ALLF, 1976).
8. NEPB, *Naturvårdsverket 1967–1977 Årsbok 1977* (Stockholm: Liber/ ALLF, 1977), p. 58 f. Cf. Högström, "Air Pollution," p. 125.
9. NEPB, *Naturvårdsverkets årsbok 1970* (Stockholm: ALLF, 1971), pp. 18 ff. *Swedish Code of Statutes* 1970:621.
10. NEPB, *Naturvårdsverkets årsbok 1970* (Stockholm: ALLF, 1971), pp. 63 ff.
11. *Riksdag Record*, 1971, 145:94 ff., 103 (Mr. Wirtén, Liberal); 145:96 ff., 105 (Mrs. Mogård, Conservative); 145:92 ff., 100 f., 105 (Ingemund Bengtsson, Minister of Agriculture).
12. Committee Report JoU 1971:59; *Riksdag Record* 150:80. The committee was worried about the scarcity of low-sulfur oil, and assumed that a lowering of sulfur content would be done only within the limits of economic feasibility and technical practicability.
13. Motions 1972:555 (Conservative); 1973:1445 (Center Party).
14. Committee reports JoU 1972:27, pp. 1 f.; 1973:12, pp. 3 f. In neither case was there any floor debate on the Committee reports in the Riksdag.
15. NEPB, *Naturvårdsverkets årsbok 1973*, p. 94 f.
16. SOU 1974:101, *Begränsning av svavelutsläpp—en studie av styrmedel* (Stockholm: Ministry of Agriculture, 1974), pp. 160 f. According to this Royal Commission report, the NEPB made a provisional halt to the sulfur content regulation program up to October 1, 1974. The main part of the exemptions given during that period (97 in all) concerned exemptions from the 1 percent rule. Only in a few cases was oil with higher sulfur content than 2.5 percent allowed. The decrease in oil consumption is thought to have compensated for the higher sulfur content.
17. Royal Commission report Ds Jo 1976:2, *Mindre svavel—bättre miljö* (Stockholm: Ministry of Agriculture, 1976), pp. 1 f.
18. Ibid., chap. 3 and 5.
19. Ibid., chap. 6.
20. Ibid., chap. 7.
21. Ibid., chap. 9, esp. pp. 144 ff.

22. Cabinet proposal 1976–77:3, pp. 24 ff.
23. Motions 1975–76:1659 (Center party); 1975–76:1686 (Liberal).
24. Committee report JoU 1976–77:4, pp. 5 ff.
25. *Riksdag Record* 1976–77: 32:86, 88 (Svante Lundkvist); 32:87 f. Mr. Larsson i Borrby, Center party); 32:84 (Mr. Hovhammar, Conservative).
26. Motions 1970 I:120 and II:127 (Liberal); Committee Report LU^3 1969:39, p. 13 f. The Riksdag rejected the Liberal proposal without debate.
27. Motion 1971:27 (Liberal).
28. Specialist panel report Ds K 1971:1, *Luftföroreningar gonom bilavgaser* (Stockholm: Ministry of Transportation, 1971), English Summary, pp. 10 ff., 28 ff. The panel's proposals meant (1) that the 1968 proposals for 1975 models would be introduced one year earlier; (2) that emission standards would be strengthened every third year, and (3) that 1974–76 standards would effectively reduce hydrocarbons and nitrogen oxides, but that carbon monoxide emissions would increase after an initial dip because of the increasing number of cars. The panel estimated the costs for 1974–76 models to increase by 400–500 crowns, and for 1977 models by 1,000 crowns. The maintenance costs would increase by 100 and 200 crowns, respectively.
29. Ibid., pp. 29, 32, 34, and 41 ff.
30. Committee report JoU 1971:51, p. 3. *Riksdag Record* 1971:117, pp. 22 f. (Mr. Hanson, Center party, and Committee Chairman). The Liberals stood alone in the committee and on the floor.
31. Committee report JoU 1972:27, p. 4. Again, the Liberals stood alone.
32. *Swedish Code of Statutes* 1972:596. NEPB, *Naturvårdsverkets årsbok 1972*, p. 82 f.
33. Motion 1973:336 (Liberal); Committee report JoU 1973:12, pp. 4 ff. (Minority view of Liberals, Conservatives, and the Center party). In the 1973 election, the socialist block—Social Democrats and Leftist-Communists—gained 175 seats in the Riksdag. The bourgeois block—Liberals, Conservatives, and the Center party—also gained 175 seats. The 1973 through 1975–76 sessions witnessed several parliamentary decisions arrived at with the help of lottery.
34. Motion 1974:319 (Liberal); Committee report JoU 1974:42, p. 2 (majority view); p. 5 (minority view); *Riksdag Record* 1974:62, pp. 44 ff.
35. Motion 1976–77:521 (Social Democrat), pp. 4 ff., 19.
36. Committee report JoU 1976–77:26, pp. 10 f., 20 ff. (majority view of Center party, Liberals, and Conservatives); 27 f. (minority view of Social Democrats).
37. *Riksdag Record* 1977:124, pp. 145 f., 151 f. (Svante Lundkvist, Social Democrat).
38. Committee Directives 1977:77 (Air Pollution Control Problems Caused by Auto Emissions), issued May 5, 1977. See pp. 3 f. See also Committee Directives 1977:40 (Improvements in the Model Certification System), issued March 10, 1977.
39. Pär Nord, "Bilarnas dåliga avgasrening största hindret för ren luft," *Miljöaktuellt. Naturvårdsverkets tidning* 5 (August 1977): 6.
40. Motion 1976–77:521 (Social Democrat).

41. Motions 1970 I:120 and II:127 (Liberal); 1971:27 (Liberal); 1972:743 (Liberal). See also motion 1971:210 (Conservative). It should be noted that the specialist panel, in its 1971 report, recommended that lead-free gasoline should be available, and could be made available, for 1977 and later car models. Cf. Specialist Panel Report Ds K 1971:1, p. 213.
42. Committee Reports LU^3 1970:39; JoU 1971:51; JoU 1972:27, which pointed to the new 1972 ordinance as essentially answering the Liberal demands, and said that administrative initiatives must be awaited; p. 2.
43. Committee Report JoU 1976-77, p. 22. The majority expressed commitment to the well-established tradition of incrementalism: "As this committee has stated on so many occasions, the board's objective is to gradually eliminate the lead content in gasoline." Cf. Henrik Tandefeldt, "Sänkt blyhalt i bensinen ger liten merkostnad för bilisten," *Miljöaktuellt Naturvårdsverkets tidning* 6 (January 1978): 6. The thrust of this incrementalistic feasibility approach is best shown by the fact that in September 1977, the domestic OK oil company began marketing gasoline containing 0.15 grams of lead per liter. Evidently, neither the Product Control Board nor the new cabinet intended to increase policy aspirations beyond implementative capabilities.
44. *Riksdag Record* 1974:125, p. 109 f. (Mrs. Lundblad, Social Democrat).
45. *Riksdag Record* 1976-77:124, p. 151 (Mr. Larsson i Borrby, Center party, chairman of the Agriculture Committee).
46. See Olof Petersson, *Valundersökningar Rapport 2: Väljarna och valet 1976* (Stockholm: SCB/Liber/ALLF, 1977), chap. 5. See esp. pp. 198 ff., 214 ff. This is borne out quite clearly also in a special survey on the voters' energy policy views; see Sören Holmberg, Jörgen Westerståhl, Karl Branzén, *Väljarna och kärnkraften* (Stockholm: Liber/Publica, 1977). In the 1976 election campaign, the editorial pages mentioned the campaign issues in the following order (percent of all campaign "messages"): energy and nuclear power, 19; economic democracy, 16; economy and employment, 15; taxes, 13; . . . environmental quality, 1; see p. 38. When asked what issues they thought most important for their choice, the answers resulted in the following ranking list (percent of all surveyed): energy and nuclear power, 21; social welfare, 18; the relationship state-economy, 14; taxes, 9; economy and employment, 8; . . . environmental quality, 3; see p. 43.
47. Motions 1971:26 and 28 (Liberal; quote from mot. 28); 1971:89 (Center party); 1972:743 (Liberal); 1973:583 (Center party and Conservative); 1973:1937 (Liberal); 1974:622 and 634 (Center party); 1974:1964 (Conservative); 1974:1965 (Center party), and 1974:1966 (Liberal).
48. Committee report BeU 1970:55 (quote); SkU 1971:50, JoU 1972:27; SkU 1972:43; SkU 1973:28; SkU 1974:59, pp. 17 f. (quote).
49. Committee report 1974:59, p. 24 (minority view of the bourgeois opposition); *Riksdag Record* 1974:142, pp. 122 f., 127 f., 132 f. (Mr. Strömberg, Liberal); p. 125 (Mr. Wikner, Social Democrat); p. 138. (Roll call vote was tied at 160-160. The subsequent lottery favored the minority view.)
50. Thomas J. Anton, *Governing Greater Stockholm. A Study of Policy*

Development and System Change (Berkeley: University of California Press, 1975), pp. 174 ff.
51. For the definitions of "valence" and "position" issues, see Angus Campbell et al., *Elections and the Political Order* (New York: John Wiley & Sons, 1966), pp. 170 f.

Chapter 9

1. For a discussion of this concept, see Giandomenico Majone, "On the Notion of Political Feasibility," *European Journal of Political Research* 3 (1975): 259–74.
2. Thomas J. Anton, *Governing Greater Stockholm* (Berkeley: University of California Press, 1975), p. 181.
3. Ibid., p. 160.
4. Of course, this would have necessitated the introduction of lead-free gasoline, something Sweden probably cannot do without other European countries doing it at the same time. However, the example shows that Swedish auto manufacturers have for a long time had the technological capacity to achieve much more stringent standards than the ones prevailing in their home country but that the politicians have not pressed them to do so.
5. The literature on public policy evaluation is growing rapidly every year. I have found the following books especially valuable: Kenneth M. Dolbeare, ed., *Public Policy Evaluation* (London and Beverly Hills, Calif.: Sage Publications, 1976); Paul Nachmias, *Policy Analysis and Evaluation* (New York: St. Martin's Press, 1978); Peter H. Rossi and Walter Williams, eds., *Evaluating Social Programs. Theory, Practice, and Politics* (New York and London: Seminar Press, 1972); and Leonard S. Rutman, ed., *Evaluation Research Methods. A Basic Guide* (London and Beverly Hills, Calif.: Sage Publications, 1978).
6. Environmental Protection Agency, *National Air Quality and Emission Trends Report, 1976* (Research Triangle Park: EPA-450/1-77-002, December 1977), p. 5-1.
7. NEPB, "Swedish Reply to the OECD Ad-hoc Group of Experts on Photochemical Oxidants," December 1977; National Central Bureau of Statistics, *Statistisk Årsbok för Sverige 1977* (Stockholm: SCB, 1977), pp. 30, 193; *Bilismen i Sverige 1977* (Stockholm: AB Bilstatistik, 1977), pp. 22 ff.
8. Environmental Protection Agency, *National Air Quality and Emissions Trends Report*, p. 5-2; U.S., Dept. of Commerce, *Statistical Abstracts of the United States 1977* (Washington, D.C.: Govt. Printing Office, 1977), pp. 28, 205 f., 634.
9. The figures are taken from SOU 1974:101, *Begränsning av svavelutsläpp–en studie av styrmedel* (Stockholm: Ministry of Agriculture, 1974), pp. 71 f.; SOU 1975:98, *Miljövård i Sverige 1975–1980* (Stockholm: Ministry of Finance, 1975), pp. 22, 66; Ds Jo 1976:2, *Mindre svavel–bättre miljö* (Stockholm: Ministry of Agriculture, 1976), pp. 18 f.

10. Environmental Protection Agency, *National Air Quality and Emissions Trends Report*, p. 5–4.
11. National Central Bureau of Statistics, *Statistisk Årsbok 1977* (Stockholm: SCB. 1977), p. 142. The index for industrial production shows only a slight leveling off in 1975.
12. Environmental Protection Agency, *National Air Quality and Emissions Trends Report*, p. 5–6; U.S., Dept. of Commerce, *Statistical Abstracts 1977*, p. 577.
13. Average emissions per vehicle kilometer decreased from 36.1 grams in 1970 to 33.1 grams in 1976 (computed from sources mentioned in note 7, this chapter). The number of highway vehicles—cars, buses, and trucks—increased from 2.45 million in 1970 to 3.06 million in 1976. During the same period, the Swedish population increased by only 155,000, to 8.24 million people (cf. note 7).
14. The average miles traveled per vehicle decreased by 200 miles in 1973, and by another 500 miles in 1974. It was then at a lower level than in 1955; cf. U.S. Dept. of Commerce, *Statistical Abstracts 1977*, pp. 206, 631 ff.
15. I cannot but mention that the total number of air polluting *industrial* installations in Sweden is about 2,800; cf. SOU 1975:98, p. 58. This should be compared to the "approximately 22,000 major emitting facilities" in the United States and the "roughly 3,500 out of compliance" in 1976; cf. U.S., Congress, Senate, *Clean Air Amendments of 1976, S. Rept. 717 to accompany S. 3219*, together with minority and individual views, 94th Cong., 2d Sess., 1976, p. 34.

Bibliography

GOVERNMENT DOCUMENTS: SWEDEN

Legislative Documents

Swedish Riksdag. The *Riksdag Record*, with Appendices (Cabinet Proposals, Committee Reports).

Swedish Code of Statutes

Ministerial Documents

Royal Ministry for Foreign Affairs et al. *Air Pollution Across National Boundaries. The Impact on the Environment of Sulfur in Air and Precipitation. Sweden's Case Study for the United Nation's Conference on the Human Environment* (Stockholm: Norstedts, 1971).
_____. *Environment Protection Act. Marine Dumping Prohibition Act. With Commentaries. Information for the United Nation's Conference on the Human Environment* (Stockholm: ALLF, 1972).
_____. *Sweden's National Report to the United Nations on the Human Environment* (Stockholm: Norstedts, 1972).

Royal Commission Reports

Ministry of Agriculture. *Begränsning av svavelutsläpp—en studie av styrmedel.* SOU 1974:101 (Stockholm: 1974).
_____. *Miljövårdsforskning. Del 1: Forskningsområdet*, SOU 1967:43 (Stockholm: 1967).
_____. *Mindre svavel—bättre miljö*, Ds Jo 1976:2 (Stockholm: 1976).
Ministry of Finance, *Miljövård i Sverige 1975-1980*, SOU 1975:98 (Stockholm: 1975).
_____. *Trafikutveckling och trafikinvesteringar*, SOU 1966:69 (Stockholm: 1966).
Ministry of Industry. *Sveriges energiförsörjning. Energipolitik och organisation*, SOU 1970:13 (Stockholm: 1970).
Ministry of Justice. *Luftförorening, buller och andra immissioner*, SOU 1966:65 (Stockholm: 1966).

226 Bibliography

Ministry of Transportation. *Avgaser från bensindrivna bilar—utredning med förslag till åtgärder*, Ds K 1968:2 (Stockholm: 1968).
———. *Luftföroreningar genom bilavgaser*, Ds K 1971: 1 (Stockholm 1971).

Agency Documents

National Central Bureau of Statistics. *Statistisk Årsbok för Sverige 1968* (Stockholm: SCB 1968).
———. *Statistisk Årsbok för Sverige 1973* (Stockholm: SCB, 1973).
———. *Statistisk Årsbok för Sverige 1977* (Stockholm: SCB, 1977).
National Environment Protection Board. *Förslag till riktlinjer för emissions— begränsande åtgärder vid luftförorenande anläggningar samt PM med kommentarer till riktvärdesförslag augusti 1969* (Stockholm: NEPB, August 1969).
———. *Miljövård i Sverige. Lagstiftning, administration, forskning och anslag* (Stockholm: ALLF, 1972).
———. "Miljövårdseffekten av statsbidragen till miljövårdande atgärder" (Stockholm: NEPB, PM 488, July 1974).
———. *Naturvårdsverkets årsbok 1969* (Stockholm: SRA, 1970).
———. *Naturvårdsverkets årsbok 1970* (Stockholm: ALLF, 1971).
———. *Naturvårdsverkets årsbok 1972* (Stockholm: ALLF, 1973).
———. *Naturvårdsverkets årsbok 1973* (Stockholm: ALLF, 1974).
———. *Naturvårdsverket 1967–1977. Årsbok 1977* (Stockholm: ALLF, 1977).
———. *Riktlinjer för luftvård* (Stockholm: NEPB Publication, 1973) p. 8.
———. *Riktvärden för luftkvalitet—svaveldioxid och stoft* (Stockholm: NEPB Publication 1976), p. 8.
———. "Swedish Reply to the OECD Ad-hoc Group of Experts on Photochemical Oxidants" (Stockholm: NEPB Air Quality Department, December 1977).

GOVERNMENT DOCUMENTS: UNITED STATES

Congressional Documents

U.S., Congress. *Congressional Record* (daily edition).
U.S., Congress, House, Committee on Interstate and Foreign Commerce. *Clean Air Act of Amendments of 1976. Report 1175 to accompany H.R. 10498*, together with additional, separate, opposing, and minority views, 94th Cong., 2d Sess., 1976.
———. *Clean Air Act Amendments of 1976, Conf. Report H 1742 to accompany S. 3219*, 94th Cong., 2d Sess., 1976.
———. *Energy Supply and Environmental Coordination Act of 1974, Report 1013 to accompany H.R. 14368*, 93d Cong., 2d Sess., 1974.
———. *Energy Supply and Environmental Coordination Act of 1974, Conf. Report H 1085 to accompany H.R. 14368*, 93d Cong., 2d Sess., 1974.
U.S., Congress, Senate, Committee on Public Works. *A Legislative History of*

the Clean Air Act Amendments of 1970 (together with a section-by-section index, prepared by the Environmental Policy Division of the Congressional Research Service of the Library of Congress for the Committee on Public Works), 93d Cong., 2d Sess., 1974, vols. 1-2.

———. *Clean Air Act Amendments of 1976, Report 717 to accompany S. 3219* (together with minority and individual views), 94th Cong., 2d Sess., 1976.

———. *Energy Emergency Act, Conf. Rept. S 681 to accompany S. 2589*, 93d Cong., 2d Sess., 1974.

———. *Veto Message from the President of the United States*, returning S. 2589, The Energy Emergency Act. S. Doc. 61, 93d Cong., 2d Sess., 1974.

———. Subcommittee on Air and Water Pollution, *Hearings on the Decision of the Administrator of the Environmental Protection Agency Regarding Suspension of the 1975 Auto Emission Standards*, 93d Cong., 1st Sess., 1973, parts 1-4.

Executive Documents

National Academy of Sciences. *Report by the Committee on Motor Vehicles* (Washington, D.C.: Govt. Printing Office, 1973).

U.S. Council on Environmental Quality. *Environmental Quality. The First Annual Report of the Council on Environmental Quality.* (Washington, D.C.: Govt. Printing Office, 1970).

———. *Environmental Quality. The Second Annual Report of the Council on Environmental Quality* (Washington, D.C.: Govt. Printing Office, 1971).

———. *Environmental Quality. The Fourth Annual Report of the Council on Environmental Quality* (Washington, D.C.: Govt. Printing Office, 1973).

———. *Environmental Quality. The Fifth Annual Report of the Council on Environmental Quality* (Washington, D.C.: Govt. Printing Office, 1974).

———. *Environmental Quality. The Sixth Annual Report of the Council on Environmental Quality* (Washington, D.C.: Govt. Printing Office, 1975).

U.S., Department of Commerce. *Statistical Abstracts of the United States 1970* (Washington, D.C.: Govt. Printing Office, 1970).

———. *Statistical Abstracts of the United States 1977* (Washington, D.C.: Govt. Printing Office, 1977).

U.S., Environmental Protection Agency. *A Progress Report* (Washington, D.C.: EPA, 1972).

———. *Impact of the Fuel Shortage on Public Attitudes toward Environmental Protection* (Washington, D.C.: EPA, June 1974).

———. *National Air Quality and Emissions Trends Report 1976* (Research Triangle Park: EPA, 1977), 450/1-77-002.

———. *Progress in the Prevention and Control of Air Pollution in 1973. Report to Congress* (Washington, D.C.: EPA, 1974).

———. *The Challenge of the Environment. A Primer to EPA's Statutory Authority*, (Washington, D.C.: EPA, 1972).

———. *The First Two Years—A Review of EPA's Enforcement Program* (Springfield, Va: NTIS, February 1973).

BOOKS AND ARTICLES

Anderson, Charles W. "Comparative Policy Analysis: The Design of Measures." *Comparative Politics* 4 (October, 1971):117-31.

———. "System and Strategy in Comparative Policy Analysis: A Plea for Contextual and Experiential Knowledge." In *Perspectives on Public Policy-Making*, edited by William Gwyn and George C. Edwards III. New Orleans: Tulane University Press, 1975.

Anderson, James E. *Public Policy-Making*. New York: Praeger Publishers, 1975.

Anton, Thomas J. *Governing Greater Stockholm, A Study of Policy Development and Change*. Berkeley: University of California Press, 1975.

———. "Policy-Making and Political Culture in Sweden." *Scandinavian Political Studies* 4 (1969):82-102.

"Auto Pollution Control Deadline Postponed." *Congressional Quarterly Almanac* 1973:653-82.

Ayres, R. E., and Miller, J. E. *Citizen Suits Under the Clean Air Act*. Washington, D.C.: EPA, no date.

Bilismen i Sverige 1977. Stockholm: AB Bilstatistik, 1977.

de Boer, Connie. "The Polls: Nuclear Energy," *Public Opinion Quarterly* 41 (Fall 1977):402-11.

Burby, John F. "EPA Alive and Well, But Meeting Stiffer Resistance." *National Journal Reports* 6 (March 23, 1974):431-38.

———. "New Committee Lineups Will Shape Clean Air Act Revisions." *National Journal Reports* 7 (January 4, 1975):12-14.

Camner, P., et al. "Air Quality Criteria and Guides for Sweden in Regard to Sulfur Dioxide and Suspended Particulates." *Nordisk Hygienisk Tidskrift* (1973): Supplement 5.

Campbell, Angus, et al. *Elections and the Political Order*. New York: John Wiley & Sons, 1966.

"Clean Air Agreement." *Congressional Quarterly Weekly Report* 34 (October 2, 1976):2725.

"Clean Air Bill Cleared with Auto Emission Deadline." *Congressional Quarterly Almanac* 1970:472-86.

"Committee Action: Auto Emissions." *Congressional Quarterly Weekly Report* 33 (July 26, 1975):1638.

"Committee Consideration: Clean Air Act Amendments." *Congressional Quarterly Weekly Report* 33 (July 12, 1975):1511.

"Committee Markup: Auto Emission Standards." *Congressional Quarterly Weekly Report* 34 (March 13, 1976):574.

"Conference Outlook: Clean Air Act Amendments." *Congressional Quarterly Weekly Report* 34 (July 26, 1977):1379.

"Congress Faces Hard Choices on Clean Air Act." *Congressional Quarterly Almanac* 1975:245-50.

"Congress Votes to Delay Clean Air Standards." *Congressional Quarterly Almanac* 1974: 738-44.

Corrigan, Richard, "Congress Is Gearing Up for Major Fight on the Clean Air Act." *National Journal* 8 (June 26, 1976):901.

Crewdson, Prudence. "Clean Air Amendments Die At Session's Close." *Congressional Quarterly Weekly Report* 34 (October 9, 1976):2927-30.
_____. "Clean Air Lobbying: Non-Deterioration." *Congressional Quarterly Weekly Report* 34 (May 1, 1976):1033-38.
_____. "Congress Faces Hard Choices on Clean Air Act." *Congressional Quarterly Weekly Report* 33 (June 7, 1975):1169-75.
_____. "Senate Committee Action: Clean Air Amendments." *Congressional Quarterly Weekly Report* 34 (February 14, 1976):311.
Dolbeare, Kenneth M., ed. *Public Policy Evaluation.* London and Beverly Hills: Sage Publications, 1976.
Dunlap, Riley E., and Van Liere, Kent D. "Further Evidence of Declining Public Concern with Environmental Problems: A Research Note." *Western Sociological Quarterly* 8 (1977): 108-13.
Dye, Thomas R. *Policy Analysis. What Governments Do, and What Difference It Makes.* University, Ala.: The University of Alabama Press, 1976.
Edwards, George C., III. "Congressional Responsiveness to Public Opinion: A Policy Perspective." *Policy Studies Journal* 5 (Summer 1977):485-91.
"Energy and Environment Notes: Clean Air." *Congressional Quarterly Weekly Report* 33 (October 18, 1975):2238.
Enloe, Cynthia H. *The Politics of Pollution in a Comparative Perspective: Ecology and Power in Four Nations.* New York: David McKay Co., 1975.
Erskine, Hazel G. "The Polls: Pollution and Its Costs." *Public Opinion Quarterly* 36 (Spring 1972):120-35.
_____. "The Polls: Pollution and Industry." *Public Opinion Quarterly* 36 (Summer 1972):263-80.
Esposito, John C. *Vanishing Air.* New York: Grossman Publishers, 1970.
Gerhardt, Paul H. "An Approach to the Estimation of Economic Losses Due to Air Pollution." Washington, D.C.: NAPCA, 1968.
Grumm, John G., "The Analysis of Policy Impact." In *The Handbook of Political Science,* vol. 6, edited by Fred I. Greenstein and Nelson W. Polsby. Reading, Mass.: Addison-Wesley Publishing Co., 1975.
Gustavsson, Sverker, "Types of Policy, Types of Politics." Unpublished manuscript.
Hancock, Donald M. *Sweden. The Politics of Post-Industrial Change.* Hinsdale, Ill.: Holt, Rinehart & Winston, Dryden Press, 1972.
Heard, Jamie. "Washington Pressures/Friends of the Earth Give Environment Interests an Activist Voice." *National Journal* 2 (1970): 1711-18.
Heidenheimer, Arnold; Heclo, Hugh; and Adams, Elisabeth Teich. *Comparative Public Policy. The Politics of Social Choice in Europe and America.* New York: St. Martin's Press, 1975.
Hofferbert, Richard I. *The Study of Public Policy.* Indianapolis: The Bobbs-Merrill Co., 1974.
Högström, Ulf. "Air Pollution in Swedish Communities." *Ambio* 4 (1975):120-25.
Holmberg, Sören; Westerståhl, Jörgen; and Branzén, Karl. *Väljarna och kärnkraften.* Stockholm: Liber/Publica, 1977.
Jones, Charles O. *Clean Air. The Policies and Politics of Pollution Control.* Pittsburgh: University of Pittsburgh Press, 1975.

———. "The Limits to Public Support: Air Pollution Agency Development." *Public Administration Review* 33 (1972):502–508.

———. "Why Congress Can't Do Policy Analysis (or words to that effect)." *Policy Analysis* 2 (1976):251–64.

Kelley, Donald R.; Wescott, R.R.; and Stunkel, K. *The Economic Superpowers and the Environment: The United States, The Soviet Union, and Japan.* San Francisco: W. H. Freeman & Company Publishers, 1976.

King, Anthony. "Ideas, Institutions, and the Policies of Government. A Comparative Analysis: Part III." *British Journal of Political Science* 3 (October 1973):409–23.

Kirschten, J. Dicken. "It's Washington Taking on Detroit in the Auto Pollution Game." *National Journal* 9 (1977):9–15.

———. "The Clean Air Conference: Something for Everybody." *National Journal* 9 (1977):1261–63.

Lijphart, Arend. "The Comparable Cases Strategy in Comparative Research." *Comparative Political Studies* 5 (1975):158–77.

Lundqvist, Lennart J. *Miljövårdsförvaltning och politisk struktur.* Lund: PRISMA/Verdandidebatt, 1971.

———. "The Comparative Study of Environmental Politics: From Garbage to Gold?" *International Journal of Environmental Studies* 11 (1978):89–97.

Magida, Arthur J. "Clean Air Act Deliberations–The Changing of the Guard." *National Journal* 8 (1976):340–41.

———. "EPA Study May Bring Reprieve for Catalytic Converter." *National Journal Reports* 7 (1975):552–58.

Majone, Giandomenico. "On the Notion of Political Feasibility." *European Journal of Political Research* 3 (1975):259–74.

Mann, Dean. "Environmental Policy." *Policy Studies Journal* 1 (1972):17–22.

Mayhew, David R. *Congress: The Electoral Connection.* New Haven: Yale University Press, 1974.

McClosky, Herbert. "Political Participation." *International Encyclopedia of the Social Sciences.* Vol. 12:252–65.

Meijer, Hans. "Bureaucracy and Policy Formulation in Sweden." *Scandinavian Political Studies* 4 (1969):103–6.

Nachmias, Paul. *Policy Analysis and Evaluation.* New York: St. Martin's Press, 1978.

Noone, James A. "Great Scrubber Debate Pits EPA Against Electric Utilities." *National Journal Reports* 6 (1974):1103–14.

———. "Energy Issues Threaten Recent Environmental Gains." *National Journal Reports* 6 (1974):305–308.

———. "Scientists' Study Upholds Clean Air Standards." *National Journal Reports* 6 (1974):1389–90.

Nord, Pär. "Bilarnas dåliga avgasrening största hindret för ren luft." *Miljöaktuellt—Naturvårdsverkets tidning* 5 (1977):6.

OECD Environment Directorate, "Environmental Policy of Sweden in Relation to the 'Guiding Principles': The Environment Protection Act and Related Legislation." OECD: Environment Directorate, 1973.

Persson, Göran, "Synpunkter på luftvårdslagstiftningens tillämpning i Sverige."

In *Luftvård 73. Konferens om nordisk luftvårdspolicy, Helsingfors 21–22 November 1973.* Helsingfors: NORDFORSKs miljövårdssekretariat (1974), p. 6.

―――――. "Different Approaches to Air Pollution Control." NEPB: October 1970.

Petersson, Olof. *Väljarna och valet 1976.* Stockholm: SCB, Valundersökningar: 1977, Rapport 2.

"Problems Cited in Meeting Clean Air Deadlines," *Congressional Quarterly Weekly Report* 32 (1974):1385–86.

Quarles, John. *Cleaning Up America. An Insider's View of the Environmental Protection Agency.* Boston: Houghton Mifflin Company, 1976.

Rathlesberger, James, ed. *Nixon and the Environment. The Politics of Devastation.* New York: Village Voice/Taurus, 1972.

Rosenbaum, Walter A. *The Politics of Environmental Concern.* New York: Praeger Publishers, 1973.

Rossi, Peter H., and Williams, Walter, eds. *Evaluating Social Programs. Theory, Practice, and Politics.* New York and London: Seminar Press, 1972.

Rutman, Leonard S., ed. *Evaluation Research Methods. A Basic Guide.* London and Beverly Hills, Calif.; Sage Publications, 1978.

Smelser, Neil J. "The Methodology of Comparative Analysis." In *Comparative Research Methods,* edited by Donald P. Warwick and Samuel Osherson. Englewood Cliffs, N.J.: Prentice-Hall, 1973.

Solesbury, William. "Issues and Innovations in Environmental Policy in Britain, West Germany, and California." *Policy Analysis* 2 (1976):1–38.

Tandefeldt, Henrik. "Sänkt blyhalt i bensinen ger liten merkostnad för bilisten." *Miljöaktuellt—Naturvårdsverkets tidning* 6 (1978):6.

Utton, Albert E., et al. *Natural Resources for a Democratic Society. Public Participation in Decision-Making.* Boulder, Colo.: Westview Press, 1976.

Vedung, Evert. *Det rationella politiska samtalet. Hur politiska budskap tolkas, ordnas och prövas.* Stockholm: Aldus/Bonniers, 1977.

―――――. "The Comparative Method and Its Neighbours." In *Power and Political Theory: Some European Perspectives,* edited by Brian Barry. London: Wiley, 1976.

Wagner, James R. "Clean Air Bill Extends Deadlines." *Congressional Quarterly Weekly Report* 35 (1977): 1629–32.

―――――. "Controversial Clean Air Report Filed." *Congressional Quarterly Weekly Report* 35 (1977):960–63.

―――――. "House Weakens Clean Air Agreements." *Congressional Quarterly Weekly Report* 35 (1977): 1023–30.

―――――. "President Signs Revisions in 1970 Clean Air Law." *Congressional Quarterly Weekly Report* 35 (1977):1713–18.

―――――. "Senate Adopts Clean Air Compromise." *Congressional Quarterly Weekly Report* 35 (1977):1135–37.

―――――. "Senate-House Differences Promise Tough Conference on Clean Air Amendments." *Congressional Quarterly Weekly Report* 35 (1977):1223.

Warwick, Donald P., and Osherson, Samuel. "Comparative Analysis in the

Social Sciences." In *Comparative Research Methods,* edited by Donald P. Warwick and Samuel Osherson. Englewood Cliffs, N.J.: Prentice-Hall, 1973.

Wiedemeyer, Neeltje. "The Polls: Do People Worry About the Future?" *Public Opinion Quarterly* 40 (1976):382–91.

Wildavsky, Aaron. "The Strategic Retreat on Objectives." *Policy Analysis* 2 (1976):499–526.

World Health Organization. *Air Quality Criteria and Guides for Urban Air Pollutants.* Geneva: WHO Technical Report, Serial No. 506, 1972.

Index

Acidification, 165–66
Air pollution: amount of, 104–7; costs of, 111–12
Air Quality Act of 1967, 4, 72
Air Quality Advisory Council of the NEPB, 85–86
Air quality control regions, 8, 15, 51–53, 56, 74, 136
Air quality criteria, 55–59, 161–62
Air Quality Department of the NEPB, 10, 45
Air resource management approach (ARM), 18–19, 39–40
Allen, James, 148
Ambient air quality standards, 7, 52, 56–58; responsibility for, 73–80, 93, 113, 142–45, 151, 160–62, 173, 186, 189
Anér, Kerstin, 72
Anton, Thomas, 179, 184
Auto Emission Control Ordinance of 1972, Swedish, 172
Auto emission standards, 10; Swedish approach to, 48–50; United States approach of 1970 to, 52–60; responsibility for setting United States, 74–80; and public participation, 93–100; judicial review of, 94–100, 103, 113, 118; suspension of United States deadlines for, 133–35, 151; changes in Swedish, 169–75, 191

Baker, Howard, 78, 81, 139–42
Bengtsson, Ingemund, 163
Bentsen, Lloyd, 150
Best available technology, 43

Billings, Leon, 158
Brodhead, William M., 137–38
Broyhill, James L., 140–41
Buckley, James L., 139, 148

Carbon monoxide, 10, 49, 56, 103–4, 133–34, 150
Carter, Jimmy, 140–42, 150
Catalytic converters, 133–34, and sulfate emissions, 137–39, 151
Center party, Swedish, 72, 86, 91, 164, 168–69, 172–74, 176
Centralization of power authority, 64
Citizen suits, 93–101, 187
Clean Air Act of 1963, United States, 72
Clean Air Act Amendments of 1970, United States, 5, 60, 72, 92, 107, 191–95
Clean Air Act Amendments of 1977, United States, 140–42, 150, 157, 184
Coal conversion, 144–50
Communist party, Swedish, 47
Comparative policy analysis: framework of, 22–29; problems of, 35–36
Conservative party, Swedish, 44, 47, 91, 163–64, 168–69, 172–74
Continuous controls, 146–49
Cooper, John Sherman, 77–81
Council on Environmental Quality, United States, 143

Dahlgren, Anders, 174
Decentralization of policy authority, 64
Dingell, John D., 137–41
Dole, Robert, 77–79

233

Eagleton, Thomas, 99
Earth Day of 1970, 5, 52, 58, 117
Economic Council for Europe (ECE), 170–72
Emission guidelines for stationary sources, Swedish, 9, 44–46; public participation in the setting of, 86–87; revision of, 160–62, 187, 189
Emission standards for stationary sources, United States, 9, 56–60; responsibility for, 74–76; public participation in the setting of, 93–99; judicial review of, 94, 113; suspension of deadlines for, 142–51
Emission trends: all sources, 190–91; stationary sources, 191–93; mobile sources, 193–94
Energy consumption and air pollution, 106
Energy production and air pollution, 113–14
Energy Resources Council, United States, 146
Energy Supply and Environmental Coordination Act of 1974, United States, 135, 143–45, 173, 184
Environmental Information System, Swedish, 12
Environmental policy, 24
Environment Protection Act of 1969, Swedish, 4, 69, 184
Environmental Protection Agency (EPA), United States, 8, 12–17; and policy responsibility, 79–81, 121, 132–37; and emission standards for stationary sources, 142–49; and weakening of political influence, 156, 187, 190, 195
EPA Administrator, 93–101; and 1973 suspension of auto emission deadlines, 133–34; and state implementation plans, 142–43, 152
European Community, 168
Exhaust controls, 48–50, 169–77

Fälldin, Thorbjörn, 168

Federal Energy Agency, United States, 144
Federal Power Commission, United States, 146
Ford, Gerald, 137, 146
Franchise Board of Environment Protection (FBEP), Swedish, 9–12, 15–17, and policy responsibilities, 70–71; public participation in proceedings of, 88–92

"Great scrubber debate", 146
Griffin, Robert, 59–60, 97, 115, 119, 140–42, 184
Gurney, Edward, 78–79

Hart, Gary, 138–39, 152–58
Hazardous pollutants, 52, 56, 73–74, 94
Highway Traffic Taxation Ordinance, Swedish, 179
House Committee on Interstate and Foreign Commerce, 52, 73, 135–37; Subcommittee on Public Health and Welfare, 52, 75
Hruska, Roman L., 97
Hydrocarbons, 10, 49, 56, 104, 133–34

Individual source control approach (ISC), 18–19, 39–40
Intermittent controls, 143–47

Johnson Administration, 73
Jones, Charles O., 121
Judicial review, 77–81, 94–100

Karlsson, Göran, 124
Kling, Herman, 67–71, 87–91, 111, 124

Lead, in gasoline, 50, 104, 169, 171–73, 175
Legal Advisory Council of Swedish cabinet, 70, 87–88, 91
Liberal party, Swedish, 43–48, 71–72, 92, 123–24, 163, 168–70, 172–79
Liquefied Petroleum Gas (LPG), 178–79

Lundkvist, Svante, 50, 104, 169, 171–73, 175

Major control approaches, 6–10; distributive, 39–40; regulatory, 39–40
Minister of Agriculture, Swedish, 46, 114, 163–64, 174
Minister of Justice, Swedish, 41–43
Minister of Transportation, Swedish, 48–49, 174
Morgan, Robert, 148
Muskie, Edmund, 5; and major control approach, 52–60; and allocation of policy powers, 73–80; and public participation, 97–100; and air pollution, 103–7; and technology vs. health, 108–10, 115; and policy escalation, 118–20; and auto emission deadlines, 134–42; and stationary source emission deadlines, 144–50; and strategic considerations for policy change, 151–58, 186–87

Nader, Ralph, 5, 118
National Academy of Sciences investigation, United States, 133
National Air Pollution Control Administration, United States, 74, 79, 111
National Air Pollution Control Council, Swedish, 3, 44–46, 86
National Environment Protection Board (NEPB), Swedish, 3–5; organization of, 10–17; and emission guidelines, 43–47; and policy responsibilities, 66–72; 81; board of directors of, 85; and public participation, 86–89; as defender of public interest, 90–92, 109, 123–24; and new emission guidelines, 160–62, and sulfur content in fuel oils, 162–66, 188, 190, 196
Nitrogen oxides, 10, 51, 104, 139, 142
Nixon, Richard, 5–6, 51–52, 56, 59, 73, 79, 115; and environmental opinion, 118, 135, 155

Palme, Olof, 109
Particulates, 51, 104, 161, 192
Photochemical oxidants, 56
Physical-environmental considerations, 28–29; as explanations of policy choice, 102–7
Poisons and Pesticides Board, Swedish, 50
Policy alternative, 25
Policy analysis, character of, 22–23; intentional mode of, 25, 35
Policy change, 32–35
Policy choice, 24; and rationality, 26–27; considerations preceding, 29–32; political context of, 31–35
Policy impact, 189–90
Policy implementation, 62; in the United States, 195–96; in Sweden, 196
Policy review, 62
Political considerations, 28; as explanation of policy choice, 116–25; as explanation of policy change, 151–58 (United States); 176–80 (Sweden); 183–184
Political context of choice, 31–34; and different policy choices, 119–25; and different policy changes, 154–58, 179–80; and different styles of policy making, 181–85
Preference calculus, 27–29
Preyer, Richardson, 138–40
Product Control Board, Swedish, 175
Public health committees, Swedish, 70
Public opinion: 1970 United States, 117–18; 1969 Swedish, 120–21; 1974–1977 United States, 153–55, 186–87; 1974–1977 Swedish, 187–88
Public participation, 16–17, 83–85, 88–92, 93–100, 187–88
Public policy, 24

Quarles, John, 152–54, 195

Randolph, Jennings, 79, 134–37, 144–46, 149

Real Estate Courts, Swedish, 88–91
Riegle, Donald W., 140–42
Rogers, Paul, 75; and auto emission deadlines, 136–40; and strategic considerations for policy change, 152–57, 186
Royal Commission on Pollution and Other Nuisances, 41–45, 65–69, 86–87, 123–24
Ruckelshaus, William, 133

Saylor, John P., 75
Secretary of Health, Education, and Welfare, 54; and policy powers, 74–79; and citizen suits, 99
Senate Committee on Public Works, 55, 75–78, 93, 97, 102–3, 137; Subcommittee on Air and Water Pollution, 55–60; 93–100
Social Democratic party, Swedish, 48–50, 71–72, 92, 122–23, 126, 163–68, 173–76, 179
Socioeconomic considerations, 28; as explanations of policy choice, 110–14
Springer, William L., 75, 80
Staggers, Harley O., 75, 108, 155
State Implementation Plans, United States, 8, 56–60, 73–80; and public participation, 93–96; suspension of deadlines for, 144–47
State Motor Vehicle Exhaust Laboratory, Swedish, 48
State Regional Boards, Swedish, 12, 88

Strategic calculus, 27–29, 31–32; and policy choice, 125–27; and changes in United States policy, 156–58; and Swedish policy developments, 177–80
Substance calculus, 27–29
Sulfur content: in fuel oils, 46–48; in coal, 144–45; Swedish regulations of, 162–69
Sulfur dioxide, 46, 104, 143–44, 162–69, 192
Swedish Association of Foundries, 3
Swedish Car Manufacturers' and Wholesale Dealers' Association, 49
Swedish Petroleum Institute, 47

Technological considerations, 28; as explanation of policy choice, 108–10
Third Legislative Committee of Swedish Riksdag, 70
Trade-off calculus, 27–29
Train, Russel, 137, 146, 152–53
Transportation, 113–14

United Auto Workers, 155–57

Volvo, 3–4, 123, 139, 153, 188

Water Act, Swedish, 42, 91
Water Courts, Swedish, 65–68, 91
Waxman, Henry A., 138–39
World Health Organization, 161–62
Wyman, Louis C., 135–36, 155